FOSSILS AND LIFE EVOLUTION

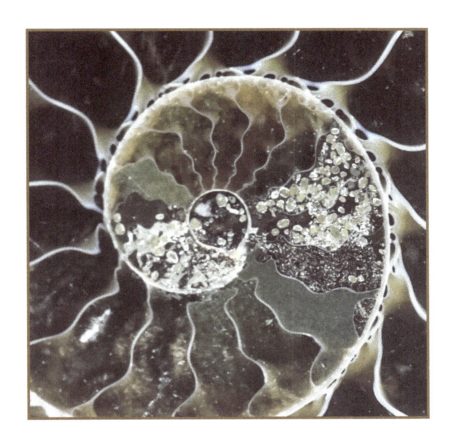

M. Dan Georgescu

Department of Geosciences, University of Calgary

Cover image © Shutterstock, Inc.

www.kendallhunt.com
Send all inquiries to:
4050 Westmark Drive
Dubuque, IA 52004-1840

Copyright © 2018 by Kendall Hunt Publishing Company

ISBN 978-1-5249-6212-8

All rights reserved. No part of this publication may be reproduced,
stored in a retrieval system, or transmitted, in any form or by any means,
electronic, mechanical, photocopying, recording, or otherwise,
without the prior written permission of the copyright owner.

Published in the United States of America

FOREWORD

Fossils and Life Evolution is a continuation of the earlier *Geological History of Life* by Georgescu and Henderson (2013) from which fragments of the text and certain illustrations were adopted. There are significant differences between *Fossils and Life Evolution* and its predecessor, and probably the most apparent is represented by the exploration of new ways to present the evolution of life forms in the Earth history to the students. *Fossils and Life Evolution* represents a new step towards the development of a new type of textbook, a long path that requires further advances in the years to come.

The author
Calgary, February 26, 2018

CONTENT

	Acknowledgments	vii
Chapter 1	**Fossils, Rocks and Geological Time**	**1**
1.1	Fossils and Their Main Classifications	1
1.2	Pre-Scientific Fossils and Beginnings of Paleontology	4
1.3	Redescovery of Fossils in the Renaissance Times	6
1.4	Paleontology and Its Subdisciplines	8
1.5	Where Fossils Can Be Found: Rocks, Sediments, Soils, and Organic Substances	10
1.6	Process of Fossilization	13
1.7	Taphonomy and Its Role in Interpreting the Fossil Record	20
1.8	Exceptional Fossil Preservation	23
1.9	Geological Time	24
1.10	Living World Hierarchy and Elements of Nomenclature	27
	Chapter Conclusions	29
Chapter 2	**Main Groups of Fossils**	**31**
2.1	Prokaryotes	31
2.2	Algae (Plant-Like Protistans)	32
2.3	Plants	33
2.4	Protozoans (Animal-Like Protistans)	36
2.5	Sponges	39
2.6	Cnidarians	40
2.7	Lophophorates	41
2.8	Worms and Worm-Like Organisms	42
2.9	Molluscs	42
2.10	Arthropods	44
2.11	Echinoderms	46
2.12	Graptolites	47
2.13	Cephalochordates and Allied Groups	48
2.14	Vertebrates	48
	Chapter Conclusions	51

Chapter 3	**Fossils and Their Applications**		**52**
	3.1	Biostratigraphy	52
	3.2	Paleobiogeography	54
	3.3	Paleoecology and Paleoenvironment Reconstructions	56
	3.4	Fossil Record Applications in the Theory of Evolution Study	57
	3.5	Fossil Uses in Economy	59
		Chapter Conclusions	62
Chapter 4	**Fossil Record and Life Evolution**		**63**
	4.1	From Biological Population to Paleontological Assemblage	63
	4.2	Intraspecific Morphologic Variability	64
	4.3	Types of Specimens in Paleontology	70
	4.4	Species in Paleontology	75
	4.5	Speciation, Species Evolution, and Extinction	77
	4.6	Principles and Methods of Species Classification	78
	4.7	Macroevolution	83
		Chapter Conclusions	85
Chapter 5	**Life Emergence and Early Evolution**		**86**
	5.1	Living Matter Composition	86
	5.2	Monomers and Polymers	89
	5.3	Datasets in Deciphering Life Emergence and Its Early Evolution	92
	5.4	Early Earth Events	92
	5.5	Bacteria, Cyanobacteria, and Stromatolites	95
	5.6	Banded Iron Formations	99
	5.7	Evolution of Eukaryotic Cells	101
	5.8	Emergence and Early Evolution of Multicellular Organisms	103
	5.9	Evolution Evolves	106
	5.10	Evolution of Burrowers and Earliest Shelly Fauna	109
	5.11	Fossils of Burgess Shale	110
		Chapter Conclusions	112
Chapter 6	**Invertebrate Evolution**		**114**
	6.1	Poriferans	114
	6.2	Cnidarians	117
	6.3	Bryozoans	121

	6.4	Brachiopods	123
	6.5	Monoplacophorans, Polyplacophorans, and Scaphopods	131
	6.6	Gastropods	133
	6.7	Bivalves	136
	6.8	Nautiloid Cephalopods and Closely Related Major Groups	140
	6.9	Ammonoid Cephalopods	143
	6.10	Coleoid Cephalopods	150
	6.11	Trilobites	152
	6.12	Crustaceans and Chelicerates	154
	6.13	Echinoderms	157
	6.14	Graptolites	163
		Chapter Conclusions	166

Chapter 7 Chordate and Vertebrate Evolution 168

	7.1	Chordates	168
	7.2	Agnathan Vertebrates	170
	7.3	Acanthodii, the First Gnathostomes	172
	7.4	Placodermi	173
	7.5	Chondrichthyes	173
	7.6	Osteichthyes	176
	7.7	Amphibians	178
	7.8	Early Reptiles (Carboniferous-Triassic)	178
	7.9	Mesozoic Aquatic Reptiles	183
	7.10	Mesozoic Flying Reptiles	185
	7.11	Dinosaurs	186
	7.12	Birds	191
	7.13	Mammals	195
		Chapter Conclusions	199

Chapter 8 Plant Evolution 201

	8.1	Colonization of Terrestrial Environments	201
	8.2	Rhyniophytes	202
	8.3	Lycophytes, Sphenophytes, and Pteridophytes	203
	8.4	Spermatophytes	205
		Chapter Conclusions	208

References 209

Index of Generic Names 212

ACKNOWLEDGMENTS

I wish to thank all the specialists who over the years have helped me in this project.

I would like to thank Dr. C. Chinery (University of Calgary) for his great help in using the paleontological collections of the University of Calgary.

Special thanks for the Museum of Natural History, Berlin for granting the permission to photograph and use specimens from their collections. I particularly thank Dr. S. Schultka (fossil plants), Dr. D. Korn (cephalopods), and Dr. F. Witzmann (fossil fishes and amphibians) for their help. I also thank Dr. L. Kluckert for the great help during this process. I thank Dr. D. Scheide for granting the permission to use photographs exhibited in the Senckenberg Natural History Museum, Frankfurt. Dr. M. Matabe is thanked for the permission to use photographs of specimens exhibited in the National Museum of Nature and Science, Tokyo. The help from Dr. K. Tanaka (Nagoya University Museum) is greatly appreciated.

I wish to express my gratitude to Dr. Hong Hua (Northwest University, Xi'an, Shaanxi), Dr. D.-G. Shu (Early Life Institute, Northwest University, Xi'an, Shaanxi) and Dr. I.I. Bucur (Babeş-Bolyai University, Cluj-Napoca) for the photographs sent and permission to use. Dr. A. Kitamura (Palaeontological Society of Japan) is thanked for the permission to use illustrations from the society's publications. The help from Dr. J. Keller (National Aeronautics and Space Administration) and C. Bridonneau (The Louvre Museum, Paris) is greatly appreciated.

I would like to thank Dr. C.M. Henderson and Dr. L. Bloom (University of Calgary) for providing excellent illustrations used in this textbook. My thanks are extended to Dr. D.K. Zelenitsky, Dr. D. Pattison, Dr. R.O. Meyer, Dr. C. Morgan (University of Calgary) and Dr. S.R. Mohr (University of Alberta) for allowing me to photograph and use rock samples and fossils from their personal collections.

Editorial boards of *Acta Palaeontologica Polonica*, *Deep Sea Drilling Project/Ocean Drilling Program*, *PLOS One*, *South African Journal of Science* and *United States Geological Survey* are thanked for allowing the use of illustrations published in their publications.

CHAPTER 1

FOSSILS, ROCKS AND GEOLOGICAL TIME

CONTENT

1.1 Fossils and Their Main Classifications
1.2 Pre-Scientific Fossils and Beginnings of Paleontology
1.3 Redescovery of Fossils in the Renaissance Times
1.4 Paleontology and Its Subdisciplines
1.5 Where Fossils Can Be Found: Rocks, Sediments, Soils, and Organic Substances
1.6 Process of Fossilization
1.7 Taphonomy and Its Role in Interpreting the Fossil Record
1.8 Exceptional Fossil Preservation
1.9 Geological Time
1.10 Living World Hierarchy and Elements of Nomenclature

Chapter Conclusions

1.1 Fossils and Their Main Classifications

Fossils are remains of older life forms, vestiges of once living organisms; the process through which a living organism is transformed into a fossil is known as *fossilization*. Not all organisms are preserved through fossilization and the ratio between the fossilized organisms and those destroyed during the process of fossilization is known as *rate of fossilization*. Not all dead organisms are considered fossils and the threshold at 11,700 years, which corresponds to the end of the last Ice Age, separates between *fossils*, which are older, and *subfossils*, which are younger. The totality of the fossils and subfossils in the Earth's crust form the *fossil record*, which is of paramount importance in reconstructing the history and evolution of life on Earth and the geological history of our planet.

2 Chapter 1 Fossils, Rocks and Geological Time

Figure 1.1 Examples of fossilized hard (1) and soft (2) body parts. 1: Specimen from the paleontological collections of the University of Calgary. 2: Specimen from the Senckenberg Natural History Museum, Frankfurt; published with permission.

Most fossils are represented by their *hard body parts* (e.g., tests, shells, valves, carapaces, etc.), which are more resistant to the process of fossilization due to their mineral nature. A coral colony (calcitic), or a trilobite carapace (chitinous) are examples of fossilized hard body parts (Figure 1.1). However, despite their resistance to the fossilization process, even hard body parts can be destroyed through processes such as dissolution, melting, etc. In contrast, *soft body parts* (e.g., tissues, organs) are only rarely preserved; their low fossilization potential is due to recycling within trophic levels and the process of organic matter decay. Soft body parts of the invertebrates can be preserved if the dead organisms are buried rapidly after death (Figure 1.1).

There are three types of fossils based according to the nature of the organism remains: *body fossils*, *trace fossils*, and *chemical fossils*. Body fossils preserve parts of the organisms itself, trace fossils represent the record of activity of a certain organism and chemical fossils are chemical substances derived from organism tissue.

- *Body fossils* represent partially or completely fossilized dead organisms; they are the most frequent kind of fossils (Figure 1.2). This category includes fossilized hard and soft body parts from all organisms (from bacteria to hominids).

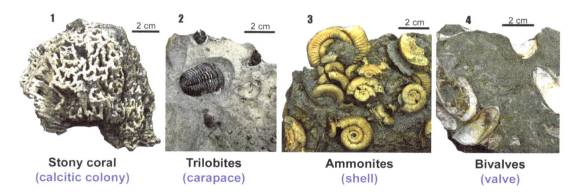

Figure 1.2 Examples of body fossils. All specimens from the paleontological collections of the University of Calgary.

Figure 1.3 Examples of trace fossils. 1 to 3: Specimens from the paleontological collections of the University of Calgary. 4: Specimen from the collections of the Museum of Natural History, Berlin; published with permission.

Most of the fossils studied in the field of paleontology are body fossils. Fossils of this kind are cab be also studied by sedimentary petrologists, stratigraphers, geochemists, etc.

- *Trace fossils* preserve organism activities, such as feeding and movement (Figure 1.3). The organisms themselves are not fossilized and clear evidence of what organism left the trails occurs rarely and they are considered cases of exceptional preservation (Figure 1.4). *Ichnology* represents the study of the trace fossils. Trace fossils are also studied by paleontologists and are commonly associated with a certain group of organisms.

- *Chemical fossil* (*biomarkers*) are chemical substances resulting from the chemical reactions between the substances or tissues in a living or dead organism and the minerals and other substances from the surrounding environment. These substances can be used as indirect evidence for the occurrence of a certain group of organisms during the geological past, although neither the organism itself, nor its traces were fossilized. Chemical fossils or biomarkers are studied by geochemists.

Mesolimulus - body fossil and trace fossil
Upper Jurassic, Solnhofen Limestone,
EU, Germany

Figure 1.4 Example of a trace fossils and the fossilized organism that produced it. Specimen from the Senckenberg Natural History Museum, Frankfurt; published with permission.

1.2 Pre-Scientific Fossils and Beginnings of Paleontology

Archaeological evidence suggests that some fossils were used for non-scientific purposes during some ancient civilizations, long before fossils started to be regarded as vestiges of ancient life forms. There is no evidence on how fossils were truly regarded in pre-scientific times but probably they were considered some curious stones. One frequently used example to demonstrate this is that ancient Egyptians used fossiliferous rocks as road milestones.

Figure 1.5 Example of pre-scientific fossil; the shell of the fossil brachiopod *Laqueus* used as ornament. Cartouche with the name of King Kheperkarê Senusret I identified by Dr. C. Bridonneau. Specimen exhibited in the Louvre Museum, Paris; photographed by the author, August-2014.

The oldest known pre-scientific fossil dates back to the Egyptian Middle Kingdom and is exhibited in the Louvre Museum in Paris (Figure 1.5). It is a brachiopod that belongs probably to the genus *Laqueus*, which bears the cartouche of Kheperkarê Senusret I, the second pharaoh of the twelfth dynasty who reigned between 1971 and 1926 B.C.; therefore, it is almost 4000 years old. This brachipod shell, which was most likely used as pendant or amulet, still bears small amounts of sediment over the surface that indicates, it is a true fossil. However, a distinct possibility is that this shell to be that of an ostreid bivalve; only the examination of the opposite side can clarify with precision if it is a brachiopod or a bivalve. Despite its importance in archaeology, there is no indication that this fossil has had any scientific significance at the time when it was inscribed with the pharaoh's name.

The first reports of fossils are known from the fragments of works remained from the scholars of the early period of the *Greek Rationalism*, which was a current of thinking that contributed to the development of the modern science. Some of the most prominent personalities of the Greek Rationalism were form what was later defined as the *Ionian School*, which flourished on the western coast of Lesser Asia, especially in the city of Miletus where the famous Thales and his disciple and follower Anaximander were active. If Thales is mostly known for his studies in mathematics, Anaximander was credited by Sagan (1985, p. 143) with the first known scientific experiment and his work *On Nature*, which did not survive to the present days, is apparently the oldest work in Greek written in prose. But the first report of fossils was given by another scholar, namely Xenophanes (~495 B.C. to ~435 B.C.), who was born in Colophon (north of Miletus), spent most of his lifetime in the Greek colonies in Sicily and southern part of Italy, and is also considered the founder of the *Eleatic School*. None of his works survived to the modern times. Although they were well-known to other scholars and philosophers in the Greek and Roman Antiquity, the only mention on the account on fossils by Xenophanes is known from the work of a much later author, Hippolytus of Rome who lived circa six centuries after him. Pease (1942) referred to Xenophanes of Colophon as the "earliest known paleontologist."

> "And Xenophanes is of opinion that there had been a mixture of the earth with the sea, and that in process of time it was disengaged from the moisture, alleging that he could produce such proofs as

the following: that in the midst of earth, and in mountains, shells are discovered; and also in Syracuse he affirms was found in the quarries the print of a fish and of seals, and in Pares an image of a laurel in the bottom of a stone, and in Melita parts of all sorts of marine animals. And he says that these were generated when all things originally were embedded in mud, and that an impression of them was dried in the mud, but that all men had perished when the earth, being precipitated into the sea, was converted into mud (…)" (translation by MacMahon 1868, p. 46–47).

Another fossil report from that period is that by Xanthus of Lydia, an enigmatical scholar who was active around the year 450 B.C. and was born in the Lydian capital Sarmis; none of his works survived to the present days. One mention of the fossil reports by Xanthus occurs in the book of the Greek geographer and historian Strabo of Amasia who lived circa four centuries later:

"Xanthus mentioned that in the reign of Artaxerxes there was so great a drought, that every river, lake, and well was dried up: and that in many places he had seen a long way from the sea fossil shells, some like cockles, other resembling scallop shells, also salt lakes in Armenia, Matiana, and Lower Phrygia, which induced him to believe that sea had formerly been where the land now was." (translation by Hamilton and Falconer 1854, p. 78).

The idea that fossils occur inland as the result of the sea diminishing lasted for nearly six centuries and it was widespread among the philosophers and artists of the early Roman Empire where we can find it, for example, in the *Metamorphoses* of the poet Publius Ovidius Naso (43 B.C. to 18 A.D.) and the essay *Isis and Osiris* on the Egyptian mythology by the historian and philosopher Plutarch of Chaeronea (~45 to ~125 A.D.). But things were about to change as we get closer to the peak of the Roman Empire when a new interpretation of the inland fossil occurrences emerged. Lucius Apuleius (~125 to ~180 A.D.) was born in Madaura (northern modern Algeria) and it is said to have been a descendant from Plutarch from his mother line. He received his early education in Carthage where he became acquainted with the Plato's philosophy, which determined him to move and continue the philosophical studies to Athens. Apuleius is mostly known as artist and many of his works survived, some in complete form; from among them the *Metamorphoses* or *The Golden Ass* is considered a masterpiece of the universal literature. In a small work titled *The Defense*, which is also known as *A Discourse on Magic*, Apuleius makes the first mention of fossil fishes from the province of Gaetulia (northern Africa).

"They say, that a woman was captivated by me through magic arts and charms derived from the sea; this, too, at a time at which they will not deny that I was far inland among the mountains of Gaetulia, where may be found fishes left by Deucalion's flood." (translation by Tighe and Gurney 1878, p. 289).

Not only this is the oldest known report of fossil fishes but it also represents a shift at paradigm level in understanding the origins of the fossils as the result of the transportation by the invading sea waters over the continents. Deucalion's flood is an event in the Greek mythology that resemble in some aspects the events described in the Mesopotamian *Epic of Gilgamesh*.

1.3 Redescovery of Fossils in the Renaissance Times

The end of the Greek Rationalism represented a major step back in European culture and science, and this period of stagnation lasted about 1000 years. Throughout the Dark Ages, fossils were either conferred a mystical role or received little attention;

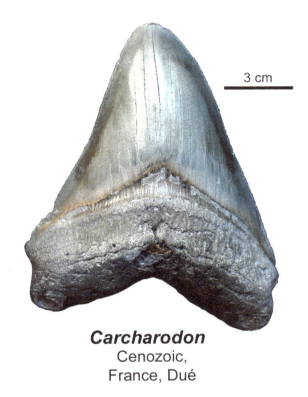

Carcharodon
Cenozoic,
France, Dué

Figure 1.6 Example of fossilized tooth of the shark genus *Carcharodon*; such fossils were referred to as glossopetrae by scholars of the late Antiquity, Dark Ages, and Renaissance. Specimen from the collections of the Museum of Natural History, Berlin; published with permission.

for example, fossilized shark teeth, which were known as *glossopetrae*, were considered woodpecker tongues (Figure 1.6).

One of the major advances during the Renaissance times happened when the fossils started to be illustrated rather than described, helping in this way at a better understanding of their organic nature, as vestiges of ancient life forms. Notably, the fossils started to be illustrated as a response shortly after Georg Bauer (1494–1555), also known by his Latin name Georgius Agricola, came up with the idea that fossils should be classified according to their shape. In time this led to the concept of *figured stones*, an idea that is nowadays abandoned by scientists, but still has supporters among philosophers.

The first illustrated fossils occur in the work *On the Metallic Objects* by Christophorus Encelius (1513–1583 and born as Christoph Entzel in Saalfield, Thuringia), which was published in Frankfurt in 1551, with a second edition from the same publisher in 1557. A variety of fossils were reported by Encelius (1551) but two of them were illustrated as woodcuts. Lanza (1984) considered the two figures represent a *Cardium* bivalve from the Plio-Pleistocene of the surroundings of Magdeburg and a gastropod, respectively (Figure 1.7). It appears evident that the text of C. Encelius is a pamphlet addressed to Georg Bauer's idea of considering fossils only according to their shape.

> "*Pliny's book 37, chapter 10, mentions what defines the chelonitids stone, that it resembles a turtle (…) where are the springs of the rivers that flow into River Alb, is a pleasant place rich in various gems, that we mentioned in the previous [part] east of the Tangra River. In that place too chelonitis gems of different species are found. At this location the first thing I discovered are chelonitis with all forms representing marine shells, sometimes perfect chelonitis, like a Jacob's shell, or marine shells, sometimes perfect chelonitis and sometimes observable with difficulty, imperfect, caught in stone,*

Figure 1.7 Earliest illustrated fossils by Encelius (1551). 1: Bivalve specimen illustrated at page 227; 2: Gastropod illustrated at page 229.

another sculpted, sometimes wonderful, sometimes perfect: in this way, altered in part, are caught in rock: both perfect and imperfect exist; sometimes big, at the size of a bean grain; sometimes smaller, firmly suggesting strength; sometimes reddish, white, shiny in the mass of marble; sometimes black lines confer the shells a marine appearance. (…) But of among the chelonitis, which resemble the marine shells, like a marine shell, *are a plenty. Besides them we have seen another chelonitis of another form that is elongate and resemble the aquatic coiled shells that can be found in rivers, as in the Alb River."* (translated from Encelius 1551, p. 227–229; underlined portions of the translation were in German and the main text in Latin).

In a next step, Conrad Gesner (1516–1565) provided the first monographic study of fossils in the work *De Rerum Fossilium* published in 1565 (Figure 1.8). It is evident that C. Gesner was influenced in his work by the concept of figured stones, but this will have little impact on the subsequent scientific studies on fossils. Less than 100 years after the publication of Gesner's work, fossils were regarded again as vestiges of ancient life forms like throughout the Antiquity times, an idea validated countless times afterwards and which represents one of the fundaments of the modern science of paleontology.

1.4 Paleontology and Its Subdisciplines

Paleontology is a branch of geology, which represents the study of fossils. Paleontology works in close connection with taphonomy (organism debris transformation between the moment of death and fossilization), paleoecology (study of the ancient ecosystems), paleobiogeography (fossil distribution in space), biostratigraphy (fossil distribution in space and time), and evolutionary paleontology (ancestor-descendant relationships between the species and higher categories) (Figure 1.9). In a broader sense paleontology is a natural science; therefore it is developed through the thorough use of the scientific method. Paleontological data are acquired through experiment and direct observation, and they are further used in interpretations. As any other of the natural sciences, paleontology is continuously changing as new and more accurate data and interpretations are produced and replace the older ones; criticism and scientific debate are two of the methods through which paleontology develops.

There are six branches of paleontology, which are primarily defined by the function of the group or groups of organisms that are studied: micropaleontology, paleoalgology, palynology, paleobotany, invertebrate paleontology, and vertebrate paleontology.

- *Micropaleontology* studies in general microscopic fossil debris; although the major group of study is represented by protistans (e.g., radiolarians, foraminifera,

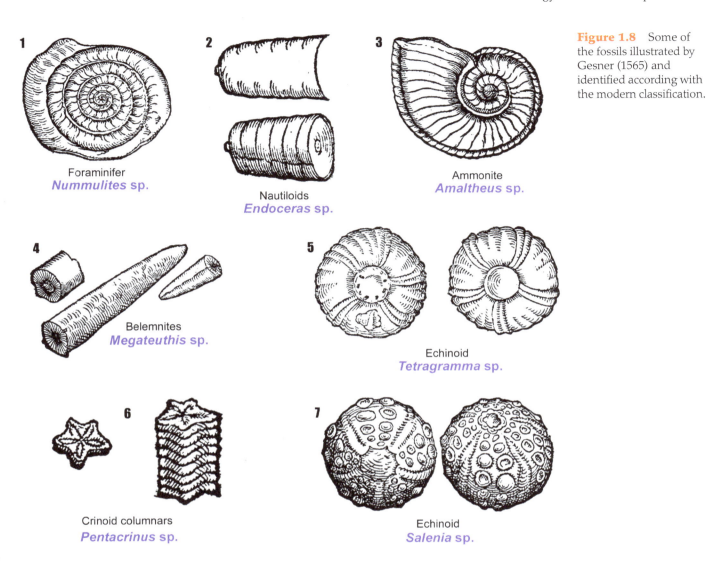

Figure 1.8 Some of the fossils illustrated by Gesner (1565) and identified according with the modern classification.

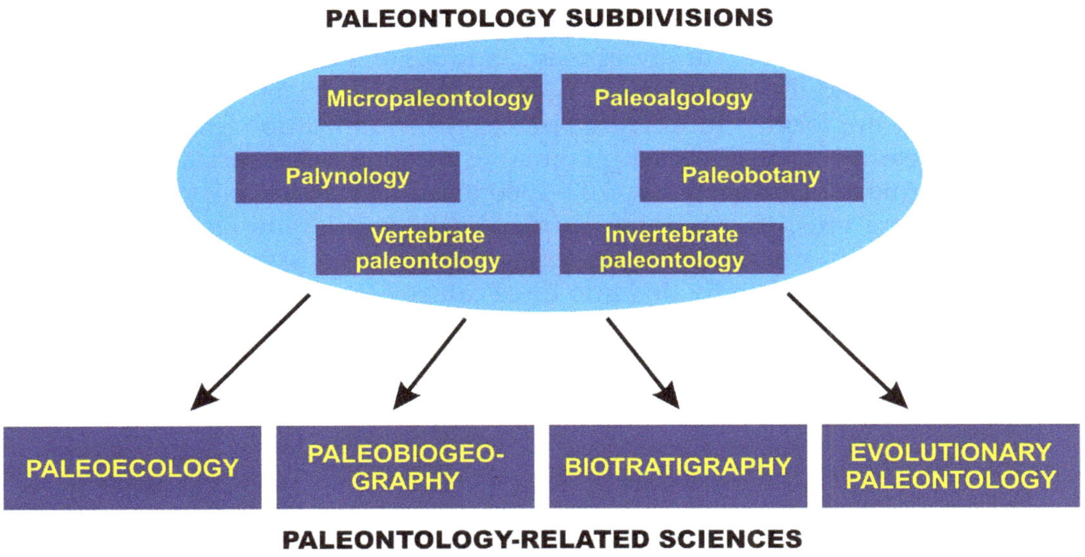

Figure 1.9 Schematic representation of the paleontology subdivisions and related sciences.

tintinids, etc.), the microscopic remains of other groups can be a topic within this branch (e.g., echinoderm plates, holoturoid sclerites, conodonts, fish teeth, etc.). Some of the species commonly studied in the field of micropaleontology can develop large-sized bodies; for example, the reefal foraminifera frequently present sizes of 1–5 centimetres and occasionally up to 12 centimetres.

- *Paleoalgology* studies the microscopic fossil algae (e.g., silicoflagellates, chlorophytes, rhodophytes, etc.); paleoalgology is considered a part of micropaleontology by many specialists.
- *Palynology* represents the study of spore and pollen grains; such microfossils are reproductive organs produced mostly by higher plants. Some other algal groups (e.g., dinoflagellates, etc.) are often considered within the range of palynology.
- *Paleobotany* is the study of the continental plant fossils, which evolved from the land colonization to present. Some algal groups (e.g., chlorophytes, charophytes) are considered by some specialists in the field of study of paleobotany.
- *Invertebrate paleontology* represents the study of the fossil multicellular animals that did not evolve a vertebral column, from the evolution of multicellularity to the present; fragments of the invertebrate tests, shells, and skeletons can be studied in the field of micropaleontology if they are of microscopic dimensions.
- *Vertebrate paleontology* studies the organisms that developed a vertebral column or its predecessor structure, the *notochord*.

1.5. Where Fossils Can Be Found: Rocks, Sediments, Soils, and Organic Substances

Most of the fossils can be found either in sediments or rocks. Rocks and sediments are aggregates of minerals, fragments of other rocks, and minerals of organic origin with a distinct, but not constant chemical and mineralogical composition. They are formed through a succession of processes and phenomena at the Earth's surface and inside its interior. The major difference between sediments and rocks is that the former are unconsolidated and the latter consist of rigidly bounded minerals. Geologists subdivide rocks, primarily based on the processes involved in their formation, into three categories known as rock families: igneous, sedimentary, and metamorphic. Each of these rock families consists of a large number of rock types. Rocks can be studied in a variety of ways, and among them the observations of hand specimens and thin sections are the most frequent.

Igneous rocks are formed through the process of solidification from molten matter deep beneath the Earth's crust (magma) or at the Earth's surface (lava). Both lava and magma are extremely hot (700–1300 °C) and as no organisms can survive at such temperatures, they completely devoid of fossils although some molds (e.g., mollusc shells, trees) may be preserved within some lava flows.

Sedimentary rocks are subdivided into three categories: detrital (clastic), chemical (precipitated), and biochemical; they form at the Earth's surface or, more rarely, in the uppermost part of the crust. They form at normal conditions of pressure and temperature, and for this reason most of the fossils on Earth are found in this kind of rocks.

- *Detrital rocks* are formed through a complex set of processes, known as the *clastic rock formation cycle*. This starts with *weathering and erosion* of pre-existing rocks under the influences of the atmospheric conditions on the continent; the pre-existing rocks can be fragmented without a change in the chemical and mineralogical composition (physical weathering), or they can have the chemical and mineralogical composition partially or completely changed (chemical weathering). The newly formed rock fragments and minerals (clasts, lithoclasts, bioclasts) are transported then toward a sedimentary basin; the *transportation process* can happen through the action of the air currents and winds (eolian transport), stream and rivers (water transport), and moving ice (glacier transport). The highest volume of sediment is transported by the continental running waters, namely streams and rivers. *Deposition* begins when the clasts reach the sedimentary basin and ends when they are settled at the basin floor; clast redistribution through bottom currents and currents are often involved during the deposition. *Lithification* is the final set of processes of the clastic rock formation cycle and its primary effect is the process of transformation of the sediments into sedimentary rocks; this involves compaction (volume reduction), cementation (mineral bounding), and recrystallization (crystal structure reorganization). Fossils can occur virtually in all detrital sediments. Detrital sedimentary rocks are primarily classified according to clast granulometry (Figure 1.10), and additional classifications take into consideration clast chemical and mineralogical composition, sphericity, roundness, presence/absence of pores in the rock mass, etc.

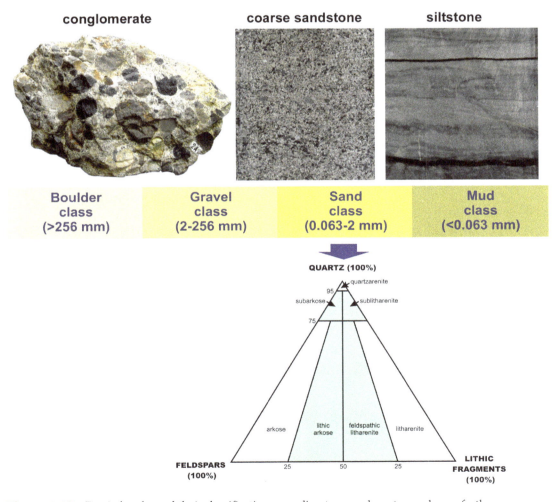

Figure 1.10 Detrital rocks and their classification according to granulometry; rocks are further subdivided according to mineralogical composition exemplified in the case of sandstones. All rock specimens from the collections of the University of Calgary.

Figure 1.11 Examples of chemically precipitated rocks. 1: Limestone. 2: Evaporites. Rock specimens from the collections of the University of Calgary.

- *Chemical rocks* form at the Earth surface or in the uppermost portion of the crust through precipitation from supersaturated solutions. Two major categories of chemical rocks can be recognized according to the chemical and mineralogical composition: carbonates and evaporites (Figure 1.11). *Carbonates* are often formed in shallow waters of tropical and subtropical regions, rarely in temperate climate; calcite ($CaCO_3$) and dolomite [$CaMg(CO_3)_2$] are the most frequent carbonate minerals. Aragonite has the same chemical composition as calcite ($CaCO_3$), but they differ in the mineral structure: the former is crystallized in the orthorhombic system, and the latter in the trigonal system; most carbonate shells in modern lakes, seas, and oceans are aragonitic. The most frequent *evaporite* minerals include halite (NaCl), gypsum ($CaSO_4 \cdot 2H_2O$), and anhydrite ($CaSO_4$); evaporites accumulate in large amounts especially in shallower regions of the continental shelves under warm and arid climatic conditions. Carbonates and evaporites occur frequently associated in lake environments and carbonate platforms. There is a high potential of fossil preservation in carbonate rocks, particularly due to the highly diverse and abundant populations of species with aragonitic and calcitic tests, shells or valves, and the frequent higher rates of sedimentation in regions where carbonate sediments accumulate. By contrast fossils are often absent from evaporite rocks; such rocks form through precipitation in environments which are highly restrictive to the vast majority of the living organisms.
- *Biochemical rocks* have the composition dominated by the hard parts (e.g., tests, shells, valves, skeletons, etc.) of ancient organisms (Figure 1.12). In general, these rocks present a relatively low diversity in chemical and mineralogical composition. Most biochemical rocks are carbonate in nature and among them the most frequent are reefal limestones and chalks. Reefal limestones are dominated by

Figure 1.12 Examples of biochemically precipitated rocks. Rock specimens from the paleontological collections of the University of Calgary.

various fossil and modern groups of organisms referred to as *reef-builders*; the reef-building organisms evolved through time and therefore reefal limestone that accumulated in different geological periods include different biotic components. Chalks are carbonate biochemical rocks formed on the deeper portions of the shelf and deeper parts of the oceanic basin through accumulation of microscopic fragments of coccolithophorids (photosynthetic algae) and foraminifera (protistans); chalk occurs in the rock record only during the last 100 million years. Siliceous biochemical rocks are formed through accumulation of small skeletal parts consisting of organic silica (SiO_2) of microscopic organisms. In this group radiolarites, diatomites, and silicoflagellithites are included; such rocks are formed through the accumulation of skeletons of radiolaria (protistans), frustules of diatoms, and silicoflagellate skeletons (algae), respectively.

A particular case of sedimentary deposits is soil, which is the regolith that covers a significant surface of continental areas. Soils are formed through the physical, chemical, and biological alteration of rocks at the Earth's surface. *Rhizoliths* (fossil roots) are frequent occurrences in fossil soils or paleosols. *Permafrost* occurs at high latitudes in the continental regions, namely in the temperate and subpolar climatic zones; large-sized vertebrates, such as mammoths, can be found fossilized in permafrost.

Metamorphic rocks are formed through the transformation of the pre-existing rocks of igneous, sedimentary or metamorphic nature, under a high temperature, pressure, and active fluid regime; during metamorphic processes the rocks remain in solid state (e.g., slate, schist, marble, eclogite, etc.) or are affected by partial melting (e.g., migmatite). The initial rock that is subject to metamorphic changes is known as *protolith*; the protolith can undergo chemical, mineralogical, or both types of changes. There are different kinds of metamorphism that occurs as a function of the position within the Earth crust and geological processes that drive the metamorphic transformation: *regional metamorphism* in plate collision zones, *contact (thermal) metamorphism* in proximity of igneous intrusions, *burial metamorphism* in large sedimentary basins where a thick pile of sediments and sedimentary rocks accumulate, and *high pressure* and *low temperature metamorphism* in regions with subduction, etc. The metamorphic rocks are transformed in various degrees during the metamorphic process, and this can be evaluated qualitatively as: low, medium, and high grade metamorphism (Figure 1.13). Fossils may be preserved in low-grade metamorphic rocks (e.g., slates, etc.), but they are also found occasionally in the medium-grade metamorphic rocks (e.g., schists); high-grade metamorphic rocks, which are intermediate between the igneous and metamorphic conditions, are completely devoid of fossil debris due to the partial melting process.

A distinct category of fossils occur in natural substances of organic origin, such as altered petroleum and amber. Petroleum forms through organic matter transformation under subsurface conditions; amber is the product of a small number of conifers. Both are highly viscous substances, which can trap various organisms and once trapped, the organisms die quickly and fossilize readily, being away from the bacterial degradation, scavengers, and atmospheric influences.

1.6 Process of Fossilization

Fossilization is the process of transformation of a dead organism into a fossil. There are only a few cases when the fossilized organism remains almost intact during and after this process; most of the organism mass is lost during the process of fossilization in the vast majority of cases. There are some cases in which the organism soft

Figure 1.13 Metamorphic rocks and their general classification according to the degree of metamorphism. Slate: courtesy of Dr. D. Pattison (University of Calgary). Schist and migmatite specimens from the collections of the University of Calgary.

parts are preserved, and such cases are considered cases of exceptional preservation. Transformation of a dead organism into a fossil is an extremely diverse process and herein the following fossilization processes are presented: permineralization, recrystallization, moldic preservation, replacement, carbonization, impregnation, fossilization through impression, congealment, dehydration, fossilization in amber, and fossilization in tar pits.

Permineralization is a frequent fossilization process in the case of organisms that present the hard body parts with internal porous structure (e.g., vertebrates, trees, etc.) (Figure 1.14). New minerals (e.g., calcite) or mineraloids (e.g., opal) are precipitated into the pore space from the subterranean fluids circulating through them. The original nature of the respective hard body part can be preserved: phosphatic (apatite) in the case of vertebrate bones, woody tissue in the case of tree trunks, etc.

Figure 1.14 Examples of fossilization through permineralization. 1: Specimen from the collections of the Museum of Natural History, Berlin; published with permission. 2: Base image courtesy of Dr. L. Bloom (University of Calgary); published with permission.

Recrystallization is one of the most frequent processes of fossilization and occurs especially among the fossils with calcitic hard body parts. The most common example of recrystallization is the transformation of aragonite (orthorhombic $CaCO_3$) into calcite (trigonal $CaCO_3$). This is particularly frequent among molluscs (e.g., gastropods, bivalves, cephalopods, scaphopods, etc.) and echinoderms (e.g., crinoids, echinoids, etc.). Recrystallization often results in the obliteration of the shell or test internal structures; therefore, the component layers with fibrous and porous ultrastructure cannot be observed anymore and the shell wall develops instead a blocky appearance. In addition, the newly formed calcite rhombohedra can be seen at the fossil surface (Figure 1.15).

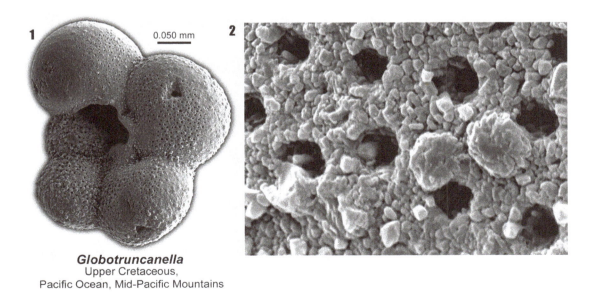

Figure 1.15 Example of recrystallization in foraminifera; original calcite is recrystallized and the newly formed calcite rhombohedra still preserve the wall high resolution morphological features.

Moldic fossilization is driven by the dissolution of the original fossil hard parts (e.g., shell, valves, test, skeleton, carapace, etc.) after the sediment around the fossil was lithified. The fossil will be partially or completely dissolved by coming in contact with chemically reactive subterranean fluids; such contact can be facilitated by the fractures in the rock and/or connected pores within the rock, namely those which allow the fluid flow in the subsurface. As a result, an empty space will be formed within the rock mass and it preserves the original fossil morphologic features (Figure 1.16). The rock around the fossil defines the *cast*, which preserves the fossil external features; rock infill forms the *mold* and it preserves the fossil internal features. Moldic fossilization is sometimes referred to as *mold-cast fossilization*.

Replacement can be considered a particular case of moldic fossilization; it happens when the empty space created within the rock through dissolution is filled with newly precipitated minerals. The process of new mineral precipitation happens from the subterranean fluids that circulate through the rock fractures and connected pores; mineral precipitation begins when the circulating fluids reach the critical concentration in a certain chemical substance. Pyrite (FeS_2) and silica (SiO_2) precipitation leads to the process of *pyritization*, which are the most frequent and spectacular cases of replacement (Figure 1.16).

Carbonization is the most common case of fossilization among plant and invertebrate fossils. During carbonization various elements of the dead body composition are expelled as organic matter decay begins; the first elements to be expelled are oxygen, nitrogen, sulfur, phosphorus, hydrogen, etc. The last element remaining from the original dead body is carbon, and this happens at a temperature in the subsurface of about 225 °C. As a result, the organism is fossilized as a thin film of graphite, a mineral with planar crystal structure, shiny black color, and greasy lustre (Figure 1.17).

Impregnation is a frequent process of fossilization when the organisms die in a basin with high concentrations of certain substances (e.g., calcium carbonate, etc.) (Figure 1.18). One classical example is that of dead algae in the tropical and subtropical regions; the original soft tissue of the algal body (thallus) becomes rigid through impregnation. Once the dead alga becomes rigid it can be fragmented by currents in

Pine cones
Lower Quaternary,
Egypt, Alexandria

Pyritized *Stropheodonta*
Lower Devonian,
USA, Oklahoma

Figure 1.16 Examples of moldic preservation (1) and replacement (2). 1: Specimen from the collection of the Museum of Natural History, Berlin; published with permission. 2: Specimen from the paleontological collections of the University of Calgary.

Figure 1.17 Examples of fossilization through carbonization in plants. 1 and 2: Specimens from the collections of the Museum of Natural History, Berlin; published with permission. 3 and 4: Specimens from the paleontological collections of the University of Calgary.

the proximity of the basin floor, and further redistributed in the basin. The process of impregnation can result in the formation of a huge number of bioclasts as the newly form fossil is broken by water currents at the bottom of the sedimentary basin.

Impressions are formed when the dead organisms are buried rapidly and their shape and sometimes internal structure are preserved on the surface of stratification despite the organic matter decay. Higher quality preservation happens if the body impression is left immediately after the organism death and before the organic matter decay begins. This fossilization process leads to the preservation of soft-bodied animals (e.g., medusas, worms, etc.), but is equally encountered in plants (e.g., leaves) (Figure 1.18).

Congealment occurs at higher latitudes of the temperate and subpolar regions that consistently present lower average temperatures over the year, and where a thick layer of *permafrost* (permanently frozen soil and rocks) is developed. The soft body parts are frequently preserved through congealment, and sometimes the preservation is so good even the last meal can be found within the stomach of the fossilized organism. A classic example of congealment is represented by the woolly mammoths of northern Siberia (Russia).

Dehydration occurs mostly in the desert areas where due to the high temperatures dead organisms lose rapidly the body fluids. The remaining parts of the dead body

Figure 1.18 Examples of fossilization through impressions (1) and impregnation (2). Both specimens are from the collections of the Museum of Natural History, Berlin; published with permission.

can fossilize if they are covered by sand. This kind of fossilization is often referred to as *mummification*. Vertebrate mummies are among the most spectacular fossils preserved this way.

Fossilization in amber results probably in the most complete preservation of the body of a dead organism. Amber is the resin produced by certain conifer species; it is a highly viscous and adherent substance and for this reason small organisms that come in contact with it can be completely enclosed in amber and fossilized (Figure 1.19). Most of the animals preserved in amber are small-sized (e.g., insects, spiders, gastropods, small frogs and salamanders, etc.). Plant debris are common occurrences in amber (e.g., grass, flowering plants, etc.). The preservation in amber is so good, that paleontologists can study even the morphology of an organism at cellular level, but fossil DNA was not recovered from such organisms. The amber producing conifers occurred relatively recently in the evolution of life on Earth; therefore, fossils preserved in amber are not known from sediments older than approximately 100 million years.

Fossilization in tar pits occurs in regions where the petroleum from the crust ascended, reached the Earth surface, and was transformed into a viscous substance

Figure 1.19 Examples of exceptional preservation: fossilization in amber (1) and tar pits (2). 1: Specimen illustrated by Dlussky and Radchenko (2009, p. 440, fig. 2B); © Acta Palaeontologica Polonica; published with permission. 2: Specimen from the paleontological collections of the University of Calgary.

(tar or bitumen) through the process of alteration, which is mostly oxidation. Animals and plants can be trapped in the tar pits and fossilize in very good conditions (Figure 1.19). A classic example is the fossil record from La Brea (California), where there is an excellent record of invertebrates (e.g., insects) and vertebrates (e.g., amphibians, reptiles, and mammals) for the last 40,000 years.

Of these types of fossilization the permineralization, recrystallization, moldic fossilization, replacement, carbonization, impregnation, and impressions occur more frequently, and are considered *common types of fossilization*. Organism preservation is of higher quality in the case of congealment, dehydration, fossilization in amber, and fossilization in tar pits; it is considered that the three-dimensional structure of the fossilized organism is preserved in the case of these four types of fossilization. In many cases the common types of fossilization co-occur resulting in *mixed fossilization*. For example, in fossilized plants the permineralization and carbonization can co-occur frequently, but in more complex situations up to four types of common fossilization can be encountered in one specimen: permineralization, carbonization, impregnation, and impressions (Figure 1.20).

Figure 1.20 Examples of mixed preservation in plants. Both specimens from the collections of the Museum of Natural History, Berlin; published with permission.

1.7 Taphonomy and Its Role in Interpreting the Fossil Record

Taphonomy studies the transformations through which a dead organism body passes through from the time of death until it is transformed into a fossil or is destroyed. Most organisms can be potentially transformed into fossil, but this process does not always happen; the vast majority of the organisms are destroyed through the process of fossilization. It is accepted in general that the *fossilization ratio* is of 1:1000 to 1:1,000,000, but frequent lower ratios occur in certain restrictive environments.

In a classical interpretation, taphonomy has two distinct components: necrolysis and biostratinomy. *Necrolysis* studies the organism dismemberment after death; its complexity largely depends of the number of hard body parts in which an organism can be broken after death. *Biostratinomy* represents the study of how organism parts are embedded within a layer; this component is often considered at the boundary between paleontology on one hand and sedimentology and sedimentary petrography on the other. Taphonomy is crucial in correctly interpreting and reconstructing the fossil record nature and the original environment in which the fossil species lived. Each high resolution paleontological study has to have a distinct taphonomy component.

Organism or fossil transportation and destruction are the most frequent processes with which paleontologists deal in current practice. The integrity of certain fossils can be readily used to assess the magnitude of fossil destruction. This assessment can be made by evaluating the *anatomical connection* of the fossil body parts (Figure 1.21). However, this method is limited by fossil morphology and cannot be applied in the case of organisms in which the hard body parts are unique or consist of a few pieces (e.g., gastropods, bivalves, cephalopods). Anatomical connection can be successfully used in the case of the fossilized organisms that have an exoskeleton consisting of a large number of pieces (e.g., echinoderms, arthropods, vertebrates, etc.). For example, the echinoderms (e.g., sea urchins, starfishes, etc.) have their bodies protected by an exoskeleton consisting of a large number of calcitic plates, which in some species are of about 10,000 in a mature specimen; exoskeleton pieces are kept in anatomical

Cactocrinus
Mississippian,
USA, Iowa

Disarticulated trilobite carapaces
Middle Cambrian,
Canada, British Columbia

Figure 1.21 Examples of fossils with the hard body parts in anatomical connection (1) and disarticulated (2). 1: Specimen from the paleontological collections of the University of Calgary. 2: Specimen courtesy Dr. S.R. Mohr (University of Alberta); published with permission.

connection for as long as the organism is alive, but the exoskeleton loses its cohesion after the organism death because of the absence of soft body parts and physiologic processes that kept the component parts together. As a result, it will disintegrate rapidly after the organism death unless is buried rapidly. Echinoderm fossils found preserving their body shape can be described as having the exoskeleton components in anatomical connection. Similarly, skeletal parts of the vertebrate bodies whenever found in anatomical connection demonstrate the absence of scavengers in the environment in which the organism lived.

Anatomical connection is in general difficult to assess quantitatively. Fossils that have the body parts in perfect anatomical connection are extremely rare; more frequently fossils have *partial anatomical connection* and frequently the fossilization process results in the complete separation of the hard body parts. Micropaleontological samples yield completely disarticulated fragments of invertebrate and vertebrate microfossils. There can be some subjectivism in assessing the anatomical connection, partial anatomical connection, and fossil disarticulation and the use of this terminology strongly depends of one specialist perspective and experience; it is a common practice among some specialists to describe fossils as being in anatomical connection although only parts of their multipiece skeleton/exoskeleton are preserved.

Taphonomic studies are extensively used to recognize ancient paleoenvironments, and especially paleobathymetry. Such estimations are determined by taking into consideration that fossils are found in layers of sedimentary rocks, which form depositional systems. A *depositional system* is a three-dimensional structure consisting of layers accumulated in a sedimentary basin under certain paleoenvironmental conditions. There are three kinds of depositional according to the zone where they formed: continental, transitional, and marine. Each depositional system is characterized by a combination of sediment source(s), relief conditions, chemical-biological factors, and succession of paleoenvironments under which sediment formation occurs. Recognizing the depositional systems is of paramount importance both in fundamental and industry-related studies; for example, such studies can be used to recognize the distribution in space and time of the source, reservoir, and seal rocks in the exploration for hydrocarbons. In addition, fossil provenance can provide information on organic matter source.

An example of industry-related taphonomic studies focuses on deep-sea turbidites, a depositional system that occurs in various forms at the slope base; the formation of deep-sea turbidites is a major process in basin filling. Deep-sea turbidites are very important in the oil-industry due to the very high porosity and permeability characteristics of the rocks in their composition (e.g., gravels, sandstones, etc.). The sediments in these depositional systems are mostly formed through submarine landslides, when vast amounts of sediments from the adjacent shelves are rapidly transported down the slope. Such sedimentation episodes and corresponding deposition of thick piles of sediments are accumulated in a short period of time, which is followed by periods of calm, when the sediments are fine-grained, resembling those formed in the deepest portions of the oceanic basin.

There are four kinds of fossils that occur in deep-sea turbidites; these groups are recognized as a function of provenance and environmental preference (Figure 1.22).

- *Reworked fossils* are much older than the deep-sea turbidites; they were fossilized before the time of deep-sea turbidite formation, therefore they are much older than the sedimentary rocks than embed them. Most of the reworked fossils originate from the sedimentary rocks in the adjacent shelf subsurface; in

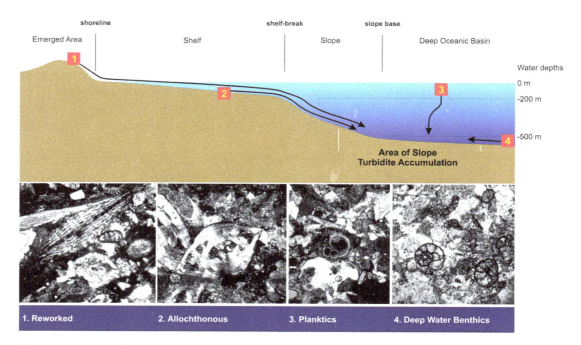

Figure 1.22 Fossil categories according to their provenance as recognized in the case of deep-sea turbidites.

cases of water currents that transport sediment from the continental regions around the basin, reworked fossils can be transported across the shelf area, before deposition at the slope base. Reworked fossils do not indicate the deep-sea turbidite age and paleobathymetry, but they help in understanding the area of sediment provenance.

- *Allochthonous fossils* are those that originally lived on shelf areas adjacent to sedimentary basins where the turbidites accumulated. They are embedded within the deep-sea turbidites during episodes when the sediments are transported down the slope. Such fossils are good age indicators, but cannot be used in estimating the deep-sea turbidite paleobathymetry.
- *Planktic fossils* develop their life cycles in the water column above the bottom area where deep-sea turbidites are formed; planktics are embedded in the turbidite sediments as they fall to the basin bottom after death. They are good age indicators, but cannot be used in paleobathymetry estimations.
- *Deep sea benthic fossils* colonize the region in which the deep-sea turbidites accumulated after the episodes of massive sediment transportation at the slope base. The environments are favourable to the sea-floor recolonization by these organisms during the periods with low water energy when only finer suspended sediments are deposited. Practically they rejuvenate the biotic component in a region of the oceanic basin that was completely sterilized by the turbidite sediment emplacement.

The ratio between the numbers of fossil specimens belonging to each of the four groups can be correlated with the deep-sea turbidite granulometry. Reworked fossils dominate in deep-sea turbidites consisting mostly of coarse sedimentary rocks (e.g., gravels, conglomerates, etc.), whereas the deep sea benthic fossils are most numerous in the fine-grained mud-dominated turbiditic systems.

1.8 Exceptional Fossil Preservation

There are cases in the fossil record where the fossil preservation is extremely good, and this is reflected especially in the occurrence of soft body parts and tests with the skeletal parts in anatomical connection. In these cases not only the burial is happened extremely fast, as the result of submarine landslides and/or high rate of sediment accumulation, but frequently the development of anoxic conditions in the proximity of the basin floor further contribute to such occurrences of exceptionally preserved fossils. In these cases, assemblages usually consisting of tens of taxa can occur in a relatively narrow geographical area, and are often called *lagerstätten* (singular: *lagerstätte*), which in German has the meaning of "places of storage" and is often translated as "fossil ores."

More than 100 lagerstätten are known from the Phanerozoic fossil record and among them are mentioned in stratigraphical order the Maotianshan Shales (Yunnan Province, China – circa 520 million years old), Burges Shale lagerstätte (British Columbia, Canada – circa 510 million years old), Hunsrück Slate (Rhineland-Palatinate, Germany – circa 400 million years old) (Figure 1.23), and Solnhofen Limestone (Bavaria, Germany – circa 155 million years old). In the cases of the lagerstätten the

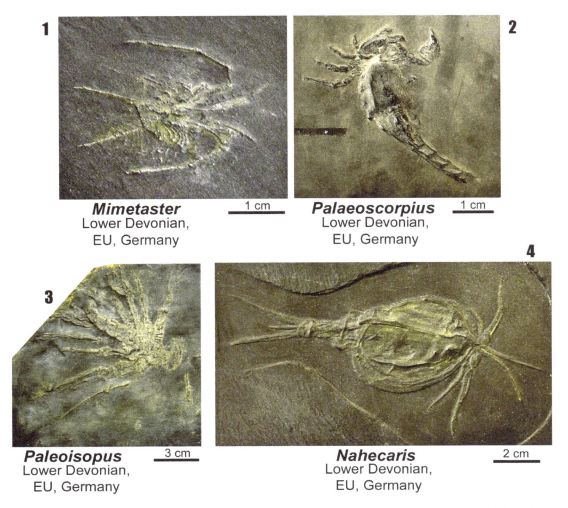

Figure 1.23 Examples of exceptionally well-preserved fossils from one lagerstätte: Hunsrück Slate of Early Devonian age from Germany. Casts from the paleontological collections of the University of Calgary.

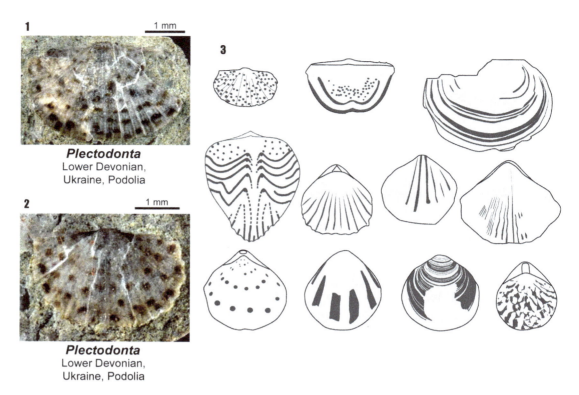

Figure 1.24 Cases of exceptional preservation in which the color patterns of brachiopod shells are fossilized. 1: Specimen from Baliński (2010, p. 697, fig. 1: A); © Acta Palaeontologica Polonica; published with permission. 2: Specimen from Baliński (2010, p. 697, fig. 1: F); © Acta Palaeontologica Polonica; published with permission. 3: Specimen from Baliński (2010, p. 699, fig. 3, simplified); © Acta Palaeontologica Polonica; published with permission.

fossilization happened through common types of fossilization. In the case of the fossilization processes that preserve the organism three-dimensional structure (congealment, dehydration, fossilization in amber, and fossilization in tar pits), practically all of them have the potential of being considered cases of exceptional preservation. One particular case of exceptional fossilization is represented by the fossilization of the color pattern of one organism; pigments are highly unstable substances that decay quickly. The color pattern is preserved in case they interact with the hard body parts shortly after death (Figure 1.24).

Cases of exceptional preservations are paramount in reconstructing the evolution of life on Earth for in general they provide a variety of data of the soft tissues, internal body organization and even data that help us reconstruct trophic chains in the case of species in which the content of the digestive tube is preserved. Fossil lagerstätten were sometimes referred to as "Polaroid pictures" of the ancient life forms on our planet.

1.9 Geological Time

Geological time is the scale used to reconstruct the succession of events in the life history on Earth. There are two scales of geological time: one relative and one absolute; the former presents a succession of events in Earth history without having numerical ages attached to it, and the latter presents the time in numerical ages.

The relative geological time scale is the result of continuous work by generations of geologists over more than 350 years. The major task was arranging the layers and bodies of rocks in the Earth crust in stratigraphic succession, namely from the older to the younger. It began with the principles of layer formation provided by Nicolaus Steno (1638–1687).

- *Principle of layer superposition*: in an undisturbed stratigraphic succession the older layer is situated at the base and the younger layers occupy a progressively higher position function of the age.
- *Principle of successive layer formation*: at the time of formation of a layer only fluid was above it.
- *Principle of original layer horizontally*: at the time of their formation the sedimentary layers were horizontal or reflect the sea-floor irregularities.
- *Principle of original layer continuity*: layers were formed over the entire surface of the basin. This principle was challenged by the subsequent studies, and especially with the development of the concept of depositional system, which showed that most depositional systems form only in certain regions of a sedimentary basin.

Subsequent studies focused on layer and facies lateral termination, and how to recognize the normal and inverse stratigraphic successions by using sedimentary structures (e.g., mudcracks, ripple-marks, burrowings, etc.). A significant leap in understanding the geological processes on Earth was made by James Hutton (1726–1784) in his book known best by the short title *Theory of the Earth* (1795). Two of Hutton's advances at paradigm level involve the geological record interpretation.

- Geological changes on Earth are gradual, and happen over long periods of time; processes such as erosion on continental areas, sediment accumulation in sedimentary basins are illustrative in this case. This idea resulted in the development of current of thinking in geology, which was later named *uniformitarianism*.
- Geological processes can be understood by studying the processes happening today on Earth. This method in understanding the rock record was later named *actualism*.

Sir William Smith (1769–1839) developed a correlation method based on the fossil content of a layer of sedimentary rocks, which further helped in ordering the succession of layers on Earth. Additional stratigraphic principles that focused on the successions of igneous rocks were given by Sir Charles Lyell (1797–1875) in *Principles of Geology*, probably the most influential geology book in the nineteenth century.

- *Principle of inclusion* postulates that the inclusions (xenoliths) are older than the rock that embeds them.
- *Principle of cross-cutting relationships* postulates that a cross-cutting rock is younger than the cross-cut rock.

All of these principles resulted in a ordering of the rock and fossil record, and for the first time it was possible to recognize a series of subdivisions in Earth history; *stratigraphy* is the branch of geology that studies the strata arrangement and succession of their formation. The history of the Earth is subdivided into four long-time intervals known as eons: Hadean, Archean, Proterozoic, and Phanerozoic; eons are further subdivided into eras (Figure 1.25).

- *Hadean* is the oldest eon in the history of our planet; it represents the time interval when our planet was a globe of molten matter. Therefore, there is no rock record for the Hadean.

26 Chapter 1 Fossils, Rocks and Geological Time

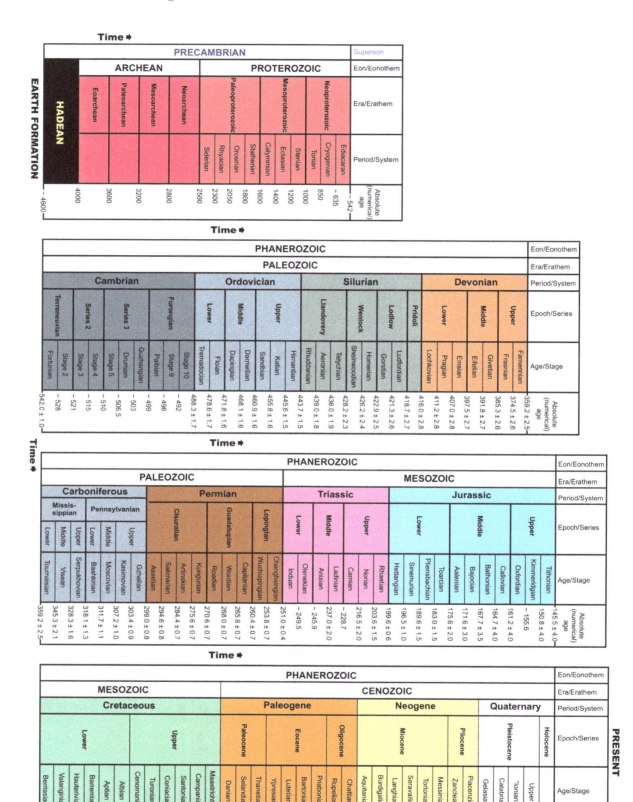

Figure 1.25 Geological time scale after Ogg and others (2008), with modifications.

- *Archean* includes the oldest rocks on Earth; the boundary between the Hadean and Archean is considered with the oldest rocks, therefore it is controlled by our knowledge on the rock record rather than rock characteristics. The Hadean/Archean boundary is 4.08 billion years in age. Archean is subdivided into four eras based on rock characteristics: Eoarchean, Paleoarchean, Mesoarchean, and Neoarchean. The oldest fossils on Earth are recorded in the Paleoarchean, but the existence of life forms can also be inferred from the Eoarchean rocks. The Archean/Proterozoic boundary is considered with the initiation of the plate tectonics.
- *Proterozoic* is the longest eon that lasted for more than two billion years. Microscopic single-celled life forms and macroscopic structures (stromatolites) produced by primitive organisms such as cyanobacteria are the dominant ones during this eon. Multicellular organisms evolved in its terminal part. The Proterozoic is subdivided into three eras according to rock and fossil record characteristics: Paleoproterozoic, Mesoproterozoic, and Neoproterozoic. The Proterozoic/Phanerozoic boundary is considered at the beginning of the processes that led to the massive development of exoskeletons.
- *Phanerozoic* is the youngest eon; the sedimentary rocks formed during this time interval frequently include large-sized forms of life. Phanerozoic eon is subdivided into three eras according to the life form resemblances to the modern ones: Paleozoic, Mesozoic, and Cenozoic. Paleozoic/Mesozoic boundary is represented by the most severe crisis in the history of life, which was generated by the ocean conveyor collapse; the boundary between the Mesozoic and Cenozoic eras is given by the meteorite impact that precipitated the dinosaur extinction along with other marine and continental species.

The eras of the Proterozoic and Phanerozoic are subdivided into periods. The periods in the Phanerozoic eras are today defined between certain bioevents, although originally they were recognized by the rock characteristics. The Phanerozoic fossil record sharply differs from those in the Archean and Proterozoic; the difference resides primarily in the large number of fossil species formally described. There is a much larger number of bioevents in the Phanerozoic when compared to the earlier eons, and for this reason more discrete stratigraphic intervals can be defined. As a result the Phanerozoic is further subdivided beyond the period level into epochs, ages and zones.

The absolute geological time scale was developed only in the twentieth century. The method is based on the process of radioactive decay, through which the nucleus of an unstable parent nucleus transforms almost spontaneously into a more stable daughter nucleus; the decay rate is constant for each pair of parent-daughter nuclei. The isotopic method is better known as radiometric dating. In the absolute geological time scale the ages are numerical. Modern technology allows most of the geological events and bioevents to be conferred an absolute (numerical) age; the measurement margin of error can follow the event or bioevent numerical age.

1.10 Living World Hierarchy and Elements of Nomenclature

Life forms classification in use today is based on the work of Carl Linnaeus (1707–1778). Linnaeus provided one classification of the organisms in the living world in the 10th edition of his work, which is known best by its shortened title, *Systema Nature*; this work was published in 1758, in a time when creationism was a norm in science. Based on Aristotelian philosophy, the life forms on Earth were grouped by Linnaeus according to the degree of similarity in a hierarchical framework with seven levels, which from the highest to the lowest are: empire, kingdom, class, order, genus, species, and variety; only five of them are retained in the modern classification (Figure 1.26). The fundamental level of living world organization is that of species. As the amount of data on various living and fossil groups increased as the result of the scientific studies, two new hierarchical levels were added to the original Linnaean framework, namely those of family (between those of genus and order) and phylum (between those of class and kingdom). This system was gradually improved, and today there are more than 30 hierarchical levels in the Linnaean classification. Hierarchical levels are formed by adding the prefix '*super-*' for the units above the basic level, and suffixes '*sub-*' and '*infra-*' for the levels below the basic level; for example, '*super*class' is above the basic level of 'class,' '*sub*class' immediately below it and '*infra*class' below subclass. An additional level, namely that of tribe was later added between the basic levels of genus and family. Taxonomy is the study of the criteria used in organism classification; the taxonomic criteria in biological and paleontological sciences are often different due to the significant differences in the data pool used by the biologists and paleontologists, respectively. The species naming system given by Linnaeus (1758) is known as the *binomial system*; each species name is formed by two names, which are in order the genus name followed by the species name. For example, genus **M** includes five species **a, b, c, d**, and **e**; the name of the five species will be **Ma, Mb, Mc, Md**, and **Me**, respectively. A more elaborate system was developed with the introduction of the *subspecies* level by Darwin (1859); the development of this new system was generated by the advances in understanding species variability. The system is known as the trinomial system, and according to it a subspecies is named by adding a third name to the binomial

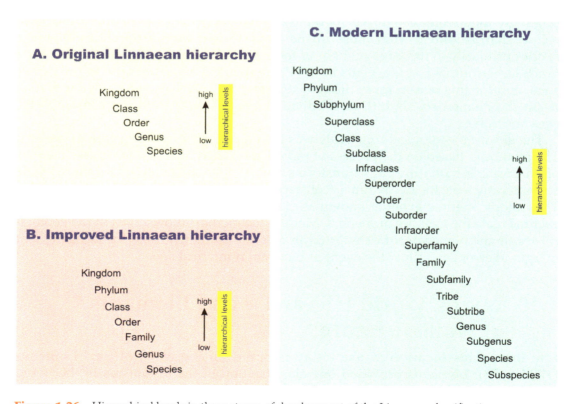

Figure 1.26 Hierarchical levels in three stages of development of the Linnaean classification.

name. For example, the species **He** includes four subspecies: **e**, **t**, **u**, and **v**; the four subspecies names are **Hee**, **Het**, **Heu**, and **Hev**. All the genus, species, and subspecies names are written in italics; this rule does not apply in the case of the higher taxonomic level units, which are given in normal font. Both systems are in use today throughout the biological and paleontological subdisciplines. All the taxonomic unit names in the Linnaean classification are given in Latin; words derived from other languages are accepted if they are Latinized. There are different terminations for certain taxonomic level (e.g., *-ida* for order, *-idae* for family, *-inae* for subfamily, etc.). Prefixes from the Ancient Greek are also accepted (e.g., *paleo*-old, *eo*-ancient, *para*-besides, *neo*-new, *eu*-true, etc.). The norms of naming taxonomic units are included in the International Code of Zoological Nomenclature (ICZN) and International Code of Botanical Nomenclature (ICBN).

The organisms of the tree of life, living, and fossil are included into five vast categories, known as kingdoms.

- *Kingdom Bacteria* includes very simple organisms, which do not have a nucleus in the cytoplasm. The oldest isolated fossils on our planet (3.465 billion years) belong to this kingdom. Bacteria dominated most of the Earth history and are preserved in stromatolites. This kingdom presents a very diverse array of metabolic strategies. The representatives of this kingdom are also referred to as *prokaryotes*.
- *Kingdom Protista* consists mostly of single-celled, solitary, and more rarely colonial, eukaryotic organisms; *eukaryotes* have well-defined nucleus or nuclei in the cytoplasm mass. Many eukaryote groups develop external protective structures (e.g., test, frustule, lorica, etc.) or an internal skeleton that has the role to support various cytoplasm components. Such structures can be fossilized relatively easily.
- *Kingdom Fungi* rarely occurs in the fossil record, although it is extremely diverse among the modern life forms; the most frequent occurrence in the fossil record is associated with fossilized plant roots (*rhizoliths*) and as *endoliths* in fossil shells.
- *Kingdom Plantae* is dominated by photosynthetic, multicellular eukaryotic organisms; higher algae and plants are included in this kingdom. Land plants evolved after the land colonization, which began circa 450 million years ago. Woody tissue is a frequent development among the continental plants, and especially in those having large-sized bodies.
- *Kingdom Animalia* includes the most evolved organisms on Earth. They are multicellular eukaryotes that present a homogeneous metabolic strategy: *aerobic heterotrophy* (*aerobic respiration*), and have the highest rate of evolutionary change among the five kingdoms. The representatives of this kingdom often present well-developed movement capabilities. The most complex animals evolved a vertebral column.

CHAPTER CONCLUSIONS

- Fossils are vestiges of ancient life forms.
- Fossilization is the process through which a living organism is transformed into a fossil.
- Not all dead organisms are considered fossils, and the boundary between fossils and subfossils is at 11,700 years; fossils are those vestiges of ancient life forms that are older than 11,700 years.

- Most of the fossils are represented by hard body parts (e.g., tests, shells, valves, carapaces, skeletons, etc.); soft body parts can be fossilized too.
- There are three kinds of fossils: body fossils, trace fossils, and chemical fossils (biomarkers).
- Paleontology is the branch of geology that represents the study of fossils; taphonomy, paleoecology, paleobiogeography, biostratigraphy, and evolutionary paleontology are paleontology-related sciences.
- Paleontology is a branch of natural sciences; it is developed through the thorough application of the scientific method.
- The organisms of the tree of life are included into five kingdoms: Bacteria, Protista, Fungi, Plantae, and Animalia.
- There are six branches of paleontology: micropaleontology, paleoalgology, palynology, paleobotany, invertebrate paleontology, and vertebrate paleontology.
- Virtually all sediments can contain fossils.
- Most of the fossils occur in the sedimentary rocks: detrital, chemical, and biochemical.
- Most of the fossils in the metamorphic realm occur in the low grade metamorphic rocks, such as slate.
- Permineralization, recrystallization, moldic preservation, replacement, carbonization, impregnation, and fossilization through impressions are considered common fossilization processes.
- Congealment, dehydration, fossilization in amber, and fossilization in tar pits preserve the original organism three-dimensional structure.
- Taphonomy is the paleontology-related science that studies the transformations through which a dead organism passes through the process of fossilization.
- Taphonomic studies are of paramount importance in reconstructing the fossil record and the original environments in which various groups of organisms developed their life cycles.
- There are four eons in Earth history: Hadean, Archean, Proterozoic, and Phanerozoic.
- The oldest rocks on Earth have an age of about 4.08 billion years.
- Geological time has two components: relative and absolute (numerical).
- Species in the fundamental level of organization of the life forms on Earth.

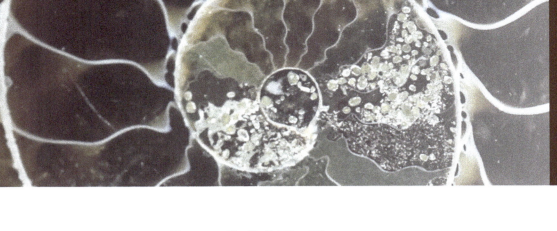

CHAPTER 2

MAIN GROUPS OF FOSSILS

CONTENT

2.1 Prokaryotes
2.2 Algae (Plant-Like Protistans)
2.3 Plants
2.4 Protozoans (Animal-Like Protistans)
2.5 Sponges
2.6 Cnidarians
2.7 Lophophorates
2.8 Worms and Worm-Like Organisms
2.9 Molluscs
2.10 Arthropods
2.11 Echinoderms
2.12 Graptolites
2.13 Cephalochordates and Allied Groups
2.14 Vertebrates

Chapter Conclusions

2.1 Prokaryotes

Prokaryotes are the most primitive organisms that occur in the fossil record; they do not have a well-defined nucleus in within cytoplasm and can occur as both solitary and colonial species. *Divisions Bacteriophyta* and *Cyanophyta* are included within the prokaryote group. They occur in the fossil record mostly as stromatolites, which are organo-sedimentary structures that dominated the life forms on Earth during the Neoarchean-Proterozoic stratigraphical interval. Stromatolitic structures

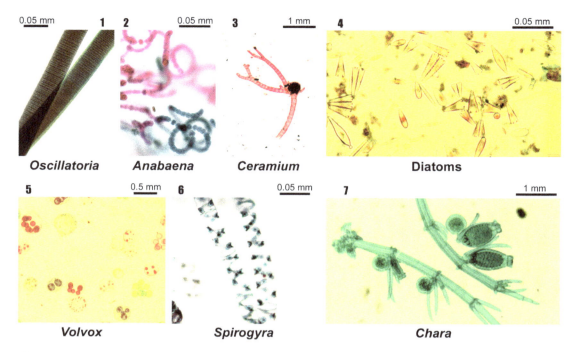

Figure 2.1 Examples of living prokaryotes (1 to 2) and plant-like protistans (3 to 7).

through the bacteria and cyanobacteria that develop their life cycles within these "menageries of microbes" shaped the atmosphere character from reducing to oxidizing and determined evolution of eukaryotes and sexual reproduction mechanism. Prokaryotes are abundant in the modern environments (Figure 2.1).

2.2 Algae (Plant-Like Protistans)

Algae are a general informal term applied to a vast number of unicellular and multicellular organisms characterized by the occurrence of one or more well-defined nuclei in the cytoplasm and plant-type metabolism (Figure 2.1). Algae are mostly autotrophic organisms with a wide range of size; in general they are <1 mm but the largest of them are >1 m in length. The representatives of this group are adapted to a variety of aquatic environments, from marine to fresh waters. Algae lack mineralized parts but their body, which is also referred to as thallus, can be impregnated with calcium carbonate after the organism death. The classification of this group is still a matter of scientific debate. Red and green algae are two of the most frequent algal divisions in the fossil record.

- *Division Rhodophyta* (red algae) are probably the oldest algal group on Earth. Most of the red algae are marine, only a relatively small number of them are adapted to brackish and fresh water environments (Figure 2.2).
- *Division Chlorophyta* (green algae) have chlorophyll within the cell plastids; these autotrophic organisms are photosynthetic and from this perspective green algae resemble the higher land plants (Figure 2.3). Chlorophytes occur in almost every aquatic environment that receives solar energy. The evolutionary occurrence of this group is in the Late Proterozoic. However, the fossil record is extremely complex and several groups are only tentatively assigned to the green algae. One of the

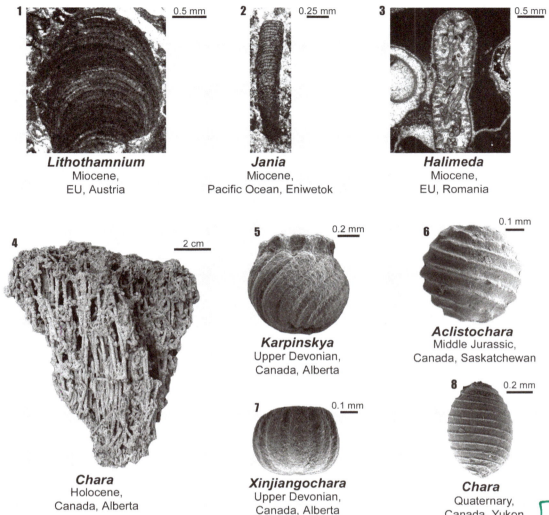

Figure 2.2 Examples of rhodophytes (1 to 2), chlorophytes (3) and charophytes (4 to 8). 2: Specimen illustrated by (Johnson, 1961, pl. 278, fig. 6); © United States Geological Survey. 3: Photograph courtesy of Dr. I.I. Bucur (Babeş-Bolyai University, Cluj-Napoca). 4: Specimen illustrated by Georgescu (2018, fig. 75). 5: Specimen illustrated by Georgescu and Braun (2006a, pl. 2, fig. 4). 6: Specimen illustrated by Georgescu and Braun (2006b, pl. 3, fig. 5). 7: Specimen illustrated by Georgescu and Braun (2006a, pl. 1, fig. 1). 8: Specimen illustrated by Georgescu (2018, fig. 79: 5).

most spectacular groups of green algae is represented by receptaculitids of the Ordovician-Permian (Figure 2.4).

- *Division Charophyta* includes morphologically advanced algae with separated sexes; the group evolved in the Early Paleozoic times (Figure 2.2). Mesozoic and Cenozoic charophytes are restricted to brackish and fresh water environments; marine species are known only from the Paleozoic record of the group.

2.3 Plants

Kingdom Plantae includes autotrophic photosynthetic organisms, which evolved in the late Silurian. The group evolved as a result of the result of an ecologic challenge, which was the invasion of terrestrial environments.

Figure 2.3 Examples of marine chlorophyte algae.

Bryophytes are the simplest land plants; they have a primitive system of transporting the nutrients and water through the plant body; this system consists of simple tubes but they lack woody tissue. Mosses are the most frequent representatives of this group in the modern floras. Reproduction is through spores.

Higher plants are considered those that developed woody tissue in the root, stalk, and branches, which allowed the plants to grow vertically and achieve larger sizes. This plant group is also known as tracheophytes; and are further subdivided into two subgroups: seedless plants and seed plants. Seedless plants have simpler morphology when compared to that of the seed plants. The following

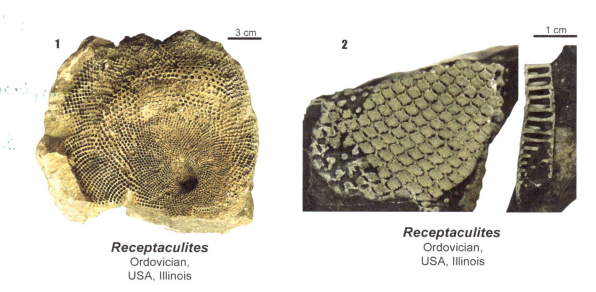

Figure 2.4 Examples of receptaculitids chlorophytes. Both specimens from the paleontological collections of the University of Calgary.

classes are included among the seedless plant group: Rhyniopsida, Lycopsida, and Pteridophyta.

- *Class Rhyniopsida* includes primitive small-sized plants that were among the earliest colonists of the terrestrial environments. Age: late Silurian-early Devonian.
- *Class Lycopsida* evolved in the late Silurian and some of its representatives exist in the modern floras. This group was the first to develop leaves. Lycopsids contributed massively to the formation of the Carboniferous coals.
- *Class Pteridophyta* includes the ferns. They evolved in Devonian and diversified several times during their evolutionary history; the highest diversification was recorded during Late Cretaceous.

Plants with seeds dominated the flora assemblages on Earth starting in the Permian. Three groups are recognized among the seed plants: pteridospermatophytes, gymnosperms, and angiosperms (Figure 2.5).

- *Class Pteridospermatophyta* is known from the Devonian-Jurassic stratigraphic interval; the representatives of this group are also known as seed ferns. The main difference when compared to the ferns resides in the reproduction mechanism, which is through seeds rather than spores.

Figure 2.5 Examples of fossilized land plants. All specimens are from the collections of the Museum of Natural History, Berlin; published with permission.

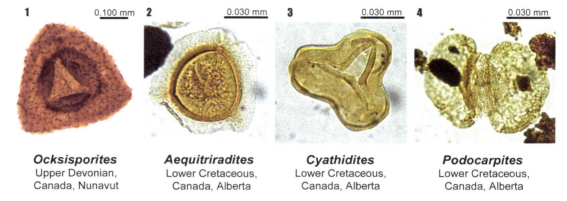

Figure 2.6 Examples of spores (1–3) and pollen (4). 1: Specimen from the L.V. Hills Collection of the University of Calgary. 2–4: Photographs courtesy of Dr. L. Bloom (University of Calgary); published with permission.

- *Class Gymnospermatophyta* evolved in the late Devonian and dominated the land floral assemblages in the Late Paleozoic and most of the Mesozoic. Gymnosperm seeds are free, not protected by fruit.
- *Class Angiospermatophyta* are the most advanced plants that evolved on Earth. The group started to dominate the land floral assemblages in the latest Cretaceous. Seeds are protected by the fruit in the representatives of this group.

Plants are known through a variety of body fossils, and practically every part of the plant can be fossilized (roots, trunks, branches, leaves, reproductive structures). Some of the most spectacular occurrences of the higher plants are the spores and pollen, which are produced by primitive and evolved higher plants, respectively (Figure 2.6). Notably, spores can be produced by other organisms such as fungi, and in these cases are referred to as *fungal spores*.

2.4 Protozoans (Animal-Like Protistans)

Protistans are a group of systematically diverse single-celled eukaryotes; many of the protistan groups have a high potential to fossilize, and for this reason the fossil record of some protistan groups is highly accurate. Foraminifera and radiolarians are examples of protistan groups.

- *Foraminifera* are heterotrophic solitary organisms, rarely colonial that protect the cytoplasm with a test, which is a protective structure of inorganic (e.g., agglutinated, calcitic, aragonitic, siliceous) or organic nature (Figure 2.7). Tests of inorganic nature can be easily fossilized. Most of the foraminiferal species are marine and only a small number of species are adapted to fresh and brackish-water environments. Foraminifera evolved in the early Cambrian but their direct ancestors are not known. The foraminiferal test grows throughout ontogeny and consists of one or more chambers; according to the number of chambers a foraminiferal test is referred to as unilocular (one chamber) or multilocular (many chambers), respectively. Most of the species of foraminifera are <1 mm in maximum dimension; frequent large-sized tests were evolved by the species living in reefal environments; development of larger tests is often associated with the increase of

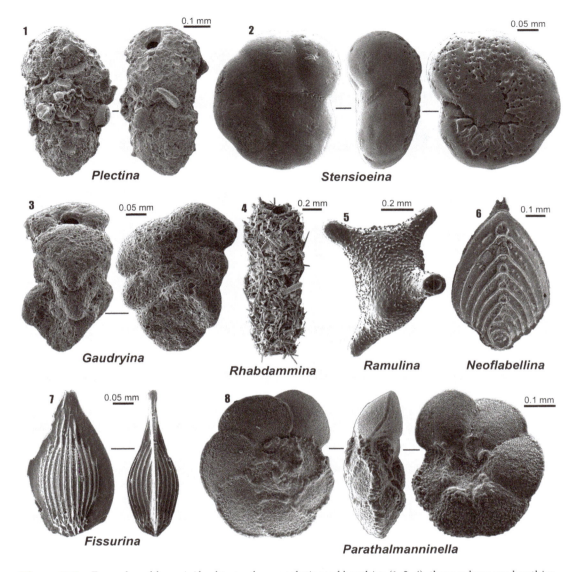

Figure 2.7 Examples of foraminiferal tests: three agglutinated benthics (1, 3–4), three calcareous benthics (2, 5–7) and one planktic (8).

number of test chambers, which in some species are more than 1000. Earlier foraminifera were benthics and only taxa with such a living mode existed throughout the Paleozoic. The group developed a planktic habitat iteratively at least three times in the Jurassic and Cretaceous; all the species of planktic foraminifera have calcitic or aragonitic tests. Foraminifera are extensively used for sediment dating and paleoenvironmental reconstructions.

- *Radiolarians* are solitary or rarely colonial protistans that present an internal skeleton consisting of strontium sulfate or organic silica; those that have the skeleton of organic silica are frequent in the fossil record and provide one of the highest quality fossil records of one group of organisms. The evolved representatives of this group are well-known among paleontologist for evolving some of the most elaborate skeleton symmetry among fossils (Figure 2.8). Radiolarians evolved in the Early Paleozoic and the earliest representatives of the group had asymmetrical skeletons or skeletons with a few elements of symmetry; the most elaborated

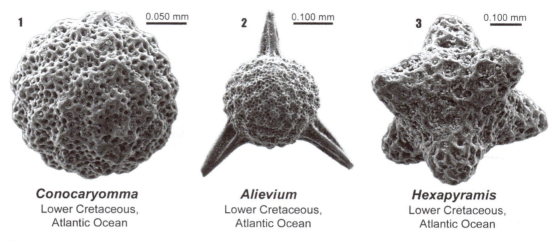

Figure 2.8 Examples of fossil radiolarian skeletons.

skeletons are known in the Mesozoic and Cenozoic times. The representatives of the radiolarian group are typical planktic organisms, which are used for the Mesozoic and Cenozoic biostratigraphy.

- *Ciliophorans* are single-celled heterotrophic solitary eukaryotes, exclusively marine that have two or more nuclei within the cytoplasm. Most of the taxa of this group protect their body with a lorica that can be organic or inorganic (Figure 2.9). Such body architecture evolved iteratively at least five times since the earliest occurrence of the group in the Ordovician times. Calpionellid ciliophorans are extensively used in the biostratigraphy of the upper Jurassic-lower Cretaceous rocks accumulated in deep marine conditions.
- *Choanoflagellates* are single-celled solitary or colonial eukaryotes with a poorly known fossil record; the only fossil choanoflagellates are known as cysts from the lower levels of the Quaternary (Figure 2.9). This group has a paramount importance in macroevolution for metazoans evolved from choanoflagellate ancestors in the Late Proterozoic.

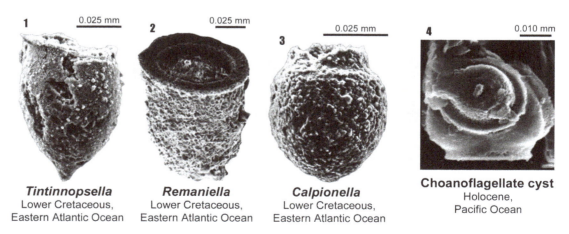

Figure 2.9 Examples of fossil ciliophorans (1–3) and choanoflagellate cyst (4). 1: Specimen illustrated by Vincent et al. (1980, pl. 8, figs 1–2); © DSDP/ODP. 2: Specimen illustrated by Luterbacher (1972, pl. 7, fig. 11); © DSDP/ODP. 3: Specimen illustrated by Vincent et al. (1980, pl. 6, fig. 5); © DSDP/ODP. 4: Specimen illustrated by Stradner and Allram (1982, fig. 1); © DSDP/ODP.

2.5 Sponges

Sponges are the simplest multicellular life forms on Earth and they are included in a distinct phylum, Porifera. Poriferans are organized in tissue grade and therefore, they did not evolve organs. The representatives of this group develop their life cycle attached to the substratum with the aid of a root-like structure and for this reason they were considered plants or polyps for a long period of time. Most of the sponge species live in marine waters but a smaller number of them adapted for fresh and brackish-water conditions. There are different types of cells in the sponge body, which are separated based on the morphology and functions they have. Scleroblasts or sclerocytes produce the spicules; spongioblasts or spongocytes produce the spongin; pinacocytes or dermal cells have the role of protection; choanocytes are flagellate cells that through the movement of their flagella create a current of water that brings in the sponge body small particles of food and dissolved oxygen used for respiration. Their size varies between 1 mm and 1.5 m in length; the largest poriferan species is *Posidonia neptuni*, which is adapted for living in reefal environments.

Sponges have in general a cylindrical or conical shape, which presents a large opening at the top (osculum) and smaller pores that penetrate through the lateral wall (Figure 2.10). The lateral wall defines a central cavity, which is also known as

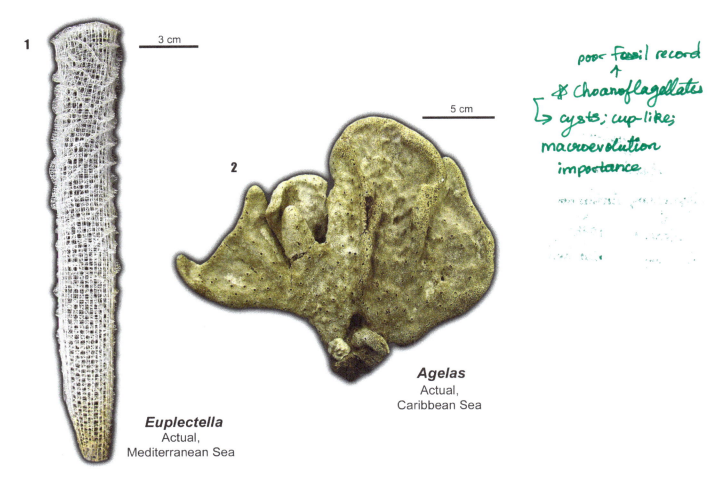

Figure 2.10 Examples of poriferans. Both specimens from the paleontological collections of the University of Calgary.

paragaster or spongocoel. This general body architecture is adapted for filter feeding; the flagellate cells (choanocytes) create a current of water that enters the body through the lateral pores and is expelled through the osculum.

An internal skeleton occurs in most of the poriferan species; the skeleton consists of small-sized spicules, which can be siliceous or calcareous. Spongin is an organic substance that is frequently associated with the siliceous spicules; relative rare species in which the skeleton consists entirely of spongin are also known. The skeleton chemical composition and architecture represent a major feature used in the poriferan systematics. Phylum Porifera is subdivided into four classes: Calcarea (calcareous skeleton), Demospongea (skeleton consists of siliceous spicules and/or spongin oriented at 60°/120°), Hexactinellida (skeleton consists of siliceous spicules with the three axes are oriented at angles of 90°) and Sclerospongia (calcareous skeleton with different skeleton architecture when compared with the representatives of class Calcarea).

The earliest evidence of the poriferan group in the fossil record is the isolated spicules of Late Proterozoic age; therefore, sponges are the oldest metazoan phylum with representatives in the modern faunas. The representatives of this phylum are relatively frequent in the fossil record, but the highest diversity and abundance occur in the periods in which the representatives of this group were reef-builders.

2.6 Cnidarians

Cnidarians are organized in tissue grade; a central cavity (gastrovascular cavity) is defined by the lateral walls and communicates with the exterior through one orifice that is used both for feeding and excretion. Cnidarians are characterized by the occurrence of nematocysts, which are cells capable to release a poison that can paralyze the prey. Another feature that separates the cnidarians from other invertebrate phyla is the alternance between a sedentary generation (polyp) and a free-swimming generation (medusa). Phylum Cnidaria is subdivided into three classes: Hydrozoa, Scyphozoa, and Anthozoa. Polyp and medusoid stages are equally developed in hydrozoans, medusa stage dominates in the scyphozoans, and polyp stage only occurs in anthozoans.

- *Class Hydrozoa* consists of small-sized marine and fresh water species and genera. The representatives of this group lack a mineralized skeleton and for this reason they rarely occur in the fossil record; the known evolutionary occurrence of the hydrozoans is in the Late Cretaceous.
- *Class Scyphozoa* includes species that have only the free-swimming medusa stage. They are not frequent in the fossil record due to the lack of a mineralized skeleton. These cnidarians are exclusively marine, some of the species having a diameter of 2 m. The evolutionary occurrence of the scyphozoans is in the Cambrian.
- *Class Anthozoa* includes most of the living and fossil cnidarians and among the modern representatives of this class are the sea pens, soft corals, and stony corals; the medusoid stage was completely lost from their growth cycle. Most species included in this class are colonial (Figure 2.11) and anthozoans were the most important reef-builders in the Earth history. The group evolved in the Cambrian.

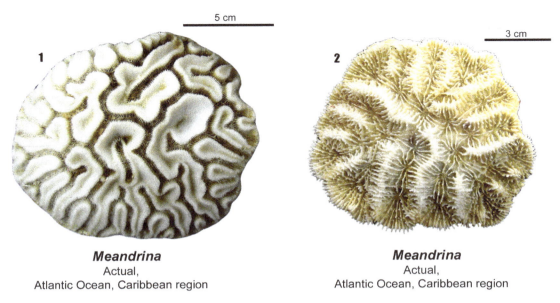

Figure 2.11 Examples of modern corals. Both specimens from the paleontological collections of the University of Calgary.

2.7 Lophophorates *very rare abundance* &most primitive organs

Lophophorates (superphylum Lophophorata) are solitary or colonial organisms, which are considered among the most primitive that evolved organs. They are characterized by the occurrence of a fan-shaped organ adapted for filter-feeding, which is known as lophophore. Most of the lophophorates are included into two phyla: Bryozoa and Brachiopoda, which evolved in the Ordovician and Cambrian times, respectively. Both groups exist in the modern faunas but are rarely abundant.

- *Phylum Bryozoa.* Bryozoans or moss animals are colonial marine organisms, and no solitary species of this phylum are known (Figure 2.12); only a small group of species are adapted to the fresh water conditions. This is the invertebrate phylum with the slowest rate to evolution.

Sponges
↳ Porifera

Sponges spicules were earliest/oldest fossil evidence available so far.

Figure 2.12 Examples of lophophorates: bryozoan (1) and brachiopods (2–3). All specimens from the paleontological collections of the University of Calgary.

- *Phylum Brachiopoda*. Together with corals and trilobites the brachiopods are among the most frequent fossils in the Paleozoic; their frequency in the Mesozoic and Cenozoic is significantly reduced. Brachiopods protect their soft body with two unequal valves, and according to the occurrence of a hinge that keeps the two valves together they are subdivided into two classes: Inarticulata and Articulata. The inarticulate brachiopods lack in general a hinge and have the two valves consisting of mixed calcium phosphate and chitin, and the valve nature in this group is often referred to as chitinophosphatic. Articulate brachiopods have well-developed teeth and sockets and the two valves are calcitic (Figure 2.12). Most of the fossil and recent brachiopods are articulates. In general brachiopods are small-sized, with a maximum dimension of several centimeters; valves considered gigantic have a length of over ten centimeters. Brachiopods were among the major reef-builders in the Late Paleozoic (Carboniferous).

2.8 Worms and Worm-Like Organisms

The informal term of "worm" is applied to a vast group of fossil and/or modern organisms that in general do not have either endoskeleton or exoskeleton; therefore, there are reduced chances for these organisms to fossilize. Fossilization is mostly as trace fossils and in this case only the traces of movement and galleries they dug in the sediments are preserved. The actual bodied of the worms and worm-like organisms are preserved only in the cases of exceptional fossilization (Figure 2.13). Phylum Platyhelminthes (flatworms) and Phylum Annelida (segmented worms) are two of the phyla included in this informal group. Annelids are rare cases among the worm group as they secret a calcitic protective structure, which can be easily fossilized; such structures are known as annelid tubes. Worm evolutionary occurrence is in the Latest Proterozoic.

2.9 Molluscs

Molluscs (Phylum Mollusca) represent a group of organisms that have the body enclosed within a soft organ known as mantle, which secrets the shell consisting of one or more pieces (Figure 2.14); there is only a relatively small number of mollusc

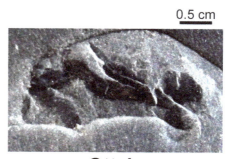

Ottoia
Middle Cambrian,
Canada, British Columbia

Figure 2.13 Example of a worm-like fossilized organism. Specimen from the paleontological collections of the University of Calgary.

2.9 Molluscs   43

Figure 2.14 Examples of mollusc hard body parts: polyplacophoran (1), gastropods (2, 5–6), bivalves (3–4, 10), scaphopods (7), and cephalopods (8–9). All specimens from the paleontological collections of the University of Calgary.

species that do not have the body protected by a shell. The main soft body has different organization plans, which are controlled primarily by the life mode of each mollusc subgroups, such as creeping, digging, active swimming; modern mollusc species do not present in general body segmentation. Molluscs evolved in the early Cambrian and are some of the most diverse organisms today in the aquatic environments. Together with the corals they were the most important reef builders throughout the Phanerozoic. Phylum Mollusca is traditionally subdivided into seven classes:

Molluscs are most diverse in aquatic systems

Aplacophora, Polyplacophora, Monoplacophora, Gastropoda, Bivalvia, Scaphopoda, and Cephalopoda. Other classes of molluscs were formalized for the representatives of this phylum of the Paleozoic.

- *Class Aplacophora* consists of mollusc organism without hard body parts; for this reason they do not occur in the fossil record. They are exclusively marine organisms.
- *Class Polyplacophora* includes molluscs which have the soft body protected by a number of plates that overlap each other. Polyplacophorans are also known under the general name of chitons. Polyplacophorans evolved in the Devonian and are rare occurrences in the fossil record.
- *Class Monoplacophora* consists of molluscs that protect the soft body parts with a shell that is conical in shape and consists of only one calcitic piece. They are exclusively marine organisms that evolved in the early Cambrian times and exist today. It seems possible that some fossil monoplacophorans to belong to the gastropod group, due to the significant morphological convergence between monoplacophorans and some gastropod taxa with uncoiled shell.
- *Class Gastropoda* consists of organisms that have the body transformed in a "foot" used for movement, and are adapted for aquatic and terrestrial environments; a small number of marine gastropods are adapted for swimming. Gastropods evolved in the Cambrian and are among the most diverse and abundant groups of organisms in the history of the Earth; they have the body protected with a twisted shell.
- *Class Cephalopoda* is characterized by the occurrence of eight or ten tentacles surrounding the mouth, and the living representatives of this class include nautiloids, octopi, and squids; this class evolved in the late Cambrian. Some cephalopods groups, such as ammonites are extensively used in the biostratigraphy of the late Paleozoic and Mesozoic sedimentary rocks. The shell is external in nautiloids and is transformed into an endoskeleton in belemnites and squids; octopi have no shell.
- *Class Scaphopoda* consists of organisms with a tapering and elongate shell that protects the soft body; the tentacles around the mouth help the animal to catch the prey. The group evolved in the Ordovician and was never among the dominant groups of molluscs.
- *Class Bivalvia* includes organisms with the body protected by two symmetrical or asymmetrical valves that are connected by a ligament, which is reinforced in many species by a hinge. Bivalves evolved in the middle Cambrian and were diverse and abundant throughout their evolutionary history. The representatives of this class do not have a head and have limited capabilities of movement. Class Rostroconcha are Paleozoic bivalve-like shells, which lack the ligament that occurs in the typical bivalves; this is a small group that became extinct at the Permian/Triassic crisis and is considered ancestral to both bivalves and scaphopods.

2.10 Arthropods

The representatives of the phylum Arthropoda are characterized by well-defined body and limb segmentation. They often have the body covered with a chitinous or calcitic protective structure, which fossilizes relatively easily, and for this reason the arthropods are frequent occurrences in the fossil record. Arthropods evolved in the early Cambrian and are abundant in the sedimentary rocks of certain stratigraphic intervals of the Phanerozoic. Evolutionary successful arthropod groups are common in the aquatic and terrestrial environments; they were among the earliest conquerors of the land circa 420 million years ago and the first invertebrates that evolved flight circa 340 million years ago; ever since a number of arthropod taxa develop a part of their life cycle in the atmosphere. Insects and spiders, for example, are among the most frequent arthropods in the modern biotopes. Being given the fact that more than one million of living species of insects populate the Earth, our times can be informally called the "Age of Insects." Phylum Arthropoda is subdivided into four subphyla: Trilobita, Chelicerata, Crustacea, and Tracheata.

- *Subphylum Trilobita* occurs in the Paleozoic (Cambrian-Permian) and became extinct at the Permian/Triassic boundary during the most severe crisis in the history of life; they reached the acme of diversity and abundance in the late Cambrian. The representatives of this subphylum have the body covered with a calcified carapace, which has a protective role. The carapace is subdivided into three parts (cephalon, thorax, and pygidium) on an anterior-posterior direction, and one axial lobe at the centre and two lateral lobes at the carapace margins on a transverse direction (Figure 2.15).

- *Subphylum Chelicerata* consists of organisms that have the body subdivided into cephalothorax (prosoma) and abdomen (opistosoma). Chelicerates are characterized by the first anterior pair of appendages which are transformed into chelicerae. The *sea scorpions* of the Paleozoic (Ordovician-Permian) were fierce predators in the shallow water marine environments and brackish waters, and some species of this group were capable to walk on the beaches; therefore they were among the earliest conquerors of land in the late Silurian, about 420 million years ago. The horseshoe crab (genus *Limulus*) of the Atlantic Canada coast is considered a "living fossil" (Figure 2.15).

- *Subphylum Crustacea*. Crustaceans are mostly aquatic organisms adapted to the marine, fresh, and brackish-water environments. Crabs, lobsters, shrimps, and barnacles are included in the crustacean group. *Ostracods* are crustaceans that have the soft body protected by two mostly unequal calcareous valves and are adapted to aquatic and more rarely terrestrial environments; the low tolerance of the ostracod

Figure 2.15 Examples of arthropods: chelicerate (1), trilobite (2), and crustaceans (3–4). All specimens from the paleontological collections of the University of Calgary.

species to environmental conditions make them excellent indicators in paleoenvironmental reconstructions (Figure 2.15).
- *Subphylum Tracheata*. Insects are the dominant group in this subphylum; insect body is subdivided into three parts: cephalon, thorax, and abdomen. One of the most spectacular achievements in the insect group was the evolution of gigantic dragonflies that could be as long as 1 m during the Carboniferous times.

2.11 Echinoderms

Echinoderms are marine and benthic organisms that develop the life cycles attached to or moving freely on the sea floor; they are solitary organisms and no cases of colonial echinoderms are known. Only a small number of species evolved a planktic mode of life. Echinoderms protect their soft body with a skeleton consisting of small calcitic pieces, which are referred to as ossicles or sclerites. Ossicles lose the connection between them after the organism death. Despite this echinoderms have a relatively well-known fossil record. Fossil taxa are found with the ossicles connected as during the life cycle only in cases of exceptional preservation. A characteristic feature of the echinoderm group is the internal water vascular system, which has the role to supplement the muscles in controlling the organism movements. Water vascular system is in connection with the tube feet, which are used for movement; tube feet are extruded through the pores of the ambulacral zones, which can be recognized at the test exterior; the zones between the ambulacral zones are referred to as interambulacral zones.

In the early Paleozoic there were echinoderm groups with few symmetry elements or that were completely asymmetrical. Pentameral symmetry is common among the evolved echinoderms. More elaborate symmetries are known in the irregular echinids where a bilateral symmetry is superimposed over the pentameral one (Figure 2.16).

Echinoderms are among the most evolved invertebrates because their internal cavity has two openings; for this reason they are included in the deuterostomate group. Their larvae resemble more the embryonic stages of the chordates rather than the larvae of the other invertebrate groups. Phylum Echinodermata is represented in the modern faunas by five classes: Ophiuroidea, Asteroidea, Holothuroidea, Crinoidea, and Echinoidea; notably, all of them evolved in the Ordovician during a period of major morphological diversification.

- *Class Ophiuroidea* (brittle stars) present five arms, which are well-separated from the rest of the body. There is no body cavity (coelomic cavity) within the arms and for this reason the arm regeneration capabilities are high.
- *Class Asteroidea* (starfishes) have resemblances in the test appearance with the ophiuroids but differ from them mainly by the occurrence of a coelomic cavity in the arms.
- *Class Holothuroidea* (sea cucumbers) have an elongate shape and are entirely benthic taxa. Sclerites are frequent in the fossil record unlike the whole holothuroid animal, which is rarely preserved.
- *Class Crinoidea* (sea lilies) are attached to the sea floor with a root-like structure, and often occur in the fossil record as detached plates. They dominated the echinoderm group in the Paleozoic but are relative rare in the Mesozoic and Cenozoic.
- *Class Echinoidea* (sea urchins) is the dominant echinoderm group in the Mesozoic and Cenozoic. In general echinoids have a globose or cordiform shape. The representatives of this group move freely on the sea floor.

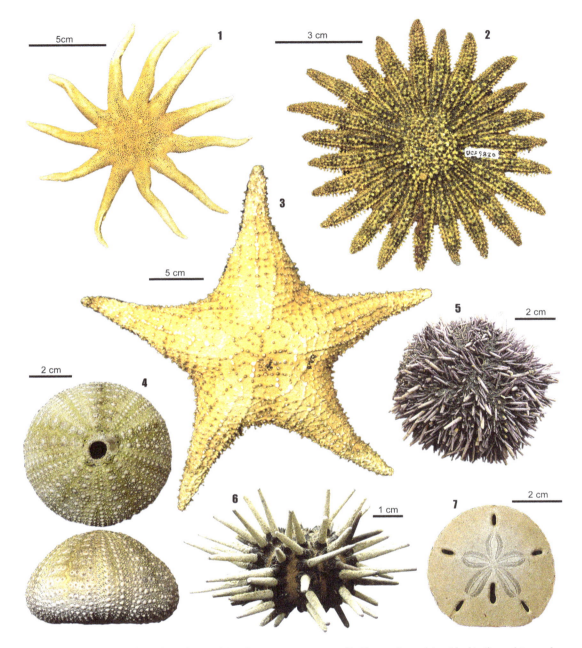

Figure 2.16 Examples of modern echinoderms: asterozoans (1–3), regular echinoids (4–6), and irregular echinoid (7). All specimens from the paleontological collections of the University of Calgary.

2.12 Graptolites

Graptolites are marine colonial organisms, which evolved both benthic and planktic habitat. The group evolved in the middle Cambrian and became extinct in the Carboniferous times; although their systematic position is debateable due to the almost complete absence of data on the soft body morphology, there is a general agreement between the paleontologists that graptolites belong to Phylum Hemichordata, where they are formalized as Class Graptolithina. Colony shape is variable: dendroid, uniserial, biserial, etc. (Figure 2.17). Graptolites are extensively used in Paleozoic biostratigraphy due to the wide distribution areas and high rates of species evolution.

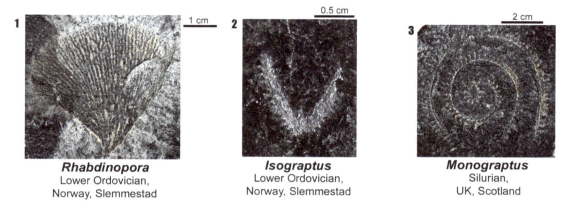

Figure 2.17 Examples of graptolites. All specimens from the paleontological collections of the University of Calgary.

The entire graptolite colony is referred to as rhabdosome; a rhabdosome is separated into one or more branches (stipes). The individuals or zooids are situated in cup-like structures known as thecae aligned along the stipes.

2.13 Cephalochordates and Allied Groups

Cephalochordates are organisms with an axial skeleton that consists of a simple rod-like structure referred to as notochord; this group of organisms occurred in the early Cambrian and exists in the modern. The complete animals are known in the fossil record only in cases of exceptional preservation; such occurrences show a strongly elongate body. Evolved from cephalochordates are the conodont animals that by contrast with their ancestors present a mineralized chewing apparatus consisting of small phosphatic pieces or teeth, which are referred to as conodont elements (Figure 2.18). Conodont animals evolved in the Late Cambrian and became extinct at the Triassic/Jurassic boundary; they provide index fossils of high accuracy over a stratigraphical interval that encompasses a time period of circa 300 million years.

2.14 Vertebrates

The most advanced life forms on Earth are included in the vertebrate group; these organisms belong to the phylum Vertebrata. The group is characterized by the axial skeleton consisting of vertebrae that are articulated forming the vertebral column. In the anterior portion of the body the vertebrates have a well-defined skull that protects the brain; an extension of the nervous system runs as spinal cord through the hollow central part of the vertebral column. The skeleton is of bony or cartilaginous nature, rarely a mixture between the two. The lower jaw is absent in the most primitive vertebrates but occurs in the evolved forms. Respiration is diverse: through gills, lungs, cutaneous, and more rarely mixed. Vertebrate classification is still a topic of scientific debate but scientists generally agree that there are six groups in the vertebrate group although the formalization is at different hierarchical levels: agnathans, fishes, amphibians, reptiles, mammals, and birds (Figure 2.19).

- *Agnathans* are vertebrates without jaws and most of these organisms are filter-feeders. Hagfishes and lampreys are modern representatives of this group, which

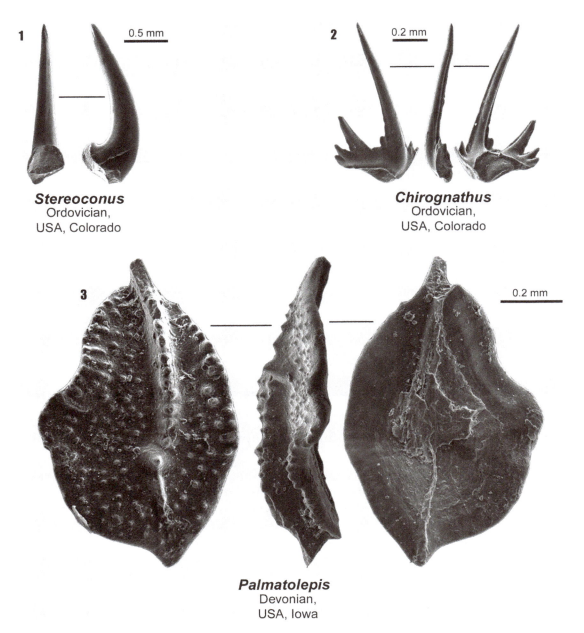

Figure 2.18 Examples of chordates of the phylum Conodonta.

dominated the vertebrate group in the Cambrian-Silurian times. Agnathans evolved in the early Cambrian probably from cephalochordate ancestors.

- *Fishes* represent one of the most successful groups of organisms in the Earth history. The axial skeleton is mostly bony but some groups evolved cartilaginous skeleton; a vestigial notochord exists is a small number of primitive species. Fishes are adapted to all aquatic environments, marine, brackish, and fresh waters; they breathe through gills and only a small number of species developed lungs. Crossopterygian fishes invaded the land in the late Silurian and they are considered the ancestors of the earliest land vertebrates.
- *Amphibians* evolved in the latest Devonian and are the most primitive land vertebrates. Respiration in the amphibian group is mixed, through lungs and cutaneous;

Figure 2.19 Examples of living vertebrates: fish (1), reptile (2), bird (3), and a mammal (4).

reproduction is through a large number of unprotected eggs. Amphibians are the ancestral group for the reptiles.

- *Reptiles* evolved in the early Carboniferous and the earliest representatives of this group were terrestrial organisms. A considerable number of Late Paleozoic (Carboniferous-Permian) reptile species developed a part of their life cycles in the aquatic environments (e.g., ponds, rivers, etc.), and reptiles fully adapted to the marine environments evolved in the Triassic times. During the Mesozoic (Jurassic-Cretaceous) the reptilian group evolved some of the largest organisms in the history of life. All living species of reptiles are cold-blooded and this was the case with many species in the group history. Some reptilian subgroups of the Permian-Cretaceous were warm-blooded and had their bodies covered with fur and feathers; evolution of flight in this group in the late Triassic apparently support this hypothesis. Reptiles dominated the terrestrial vertebrates during the Carboniferous-Cretaceous; they were dramatically reduced by the crisis produced by the meteorite impact at the Cretaceous/Paleogene boundary.

- *Mammals* occurred in the late Triassic and evolved the highest levels of intelligence in the life history on Earth; human species belongs to the mammalian group. Most mammals have the body covered with hair; they are warm-blooded and with few exceptions give birth to living offspring. Evolution of consciousness is another major achievement of the mammalian group.

- *Birds* evolved from reptilian ancestors; the lineage that will ultimately lead to birds was initiated in the late Jurassic but the earliest species with fully developed avian featured did not occur until the Cretaceous times. Birds have most of the body covered with feathers; they are warm-blooded organisms and evolved an advanced mechanism of body temperature regulation. Reproduction is through a small number of protected eggs.

CHAPTER CONCLUSIONS

- Algal groups that are the most frequent in the fossil record are the red (Rhodophyta) and green algae (Chlorophyta).
- Red algae are the oldest algal group in the fossil record.
- Plants are autotrophic photosynthetic organisms that evolved woody tissue and a vascular system of water and nutrients transport that increased in complexity throughout their evolution.
- Plant evolution began with the invasion of land during the Silurian, possibly in the Ordovician.
- Foraminifers and radiolarians are the most frequent protistans with animal-type metabolism in the fossil record.
- Sponges are the simplest multicellular organisms in the modern faunal assemblages; they do not have specialized organs.
- Cnidarians are solitary and colonial organisms organized in tissue grade; they are probably the most important reef-builders throughout Earth history.
- Worms are soft-bodied organisms with elongate body, segmented or flattened, which rarely occur in the fossil record as body fossils.
- Annelids are the most frequent worms in the fossil record.
- Lophophorates are solitary (Brachiopoda) and colonial (Bryozoa) organisms.
- Brachiopods are some of the most frequent fossils in the Paleozoic.
- Mollusc skeletal parts are extremely diverse from a morphologic point of view; mollusc species without a shell are relatively rare.
- Arthropods have segmented body and limbs.
- Trilobites are the most frequent arthropod fossils in the Paleozoic; with more than one million living species, insects are the dominant organisms on Earth today.
- Echinoderms have the body protected by a calcitic exoskeleton with elaborated symmetry but some taxa are completely asymmetrical; they have one of the highest diversity in the body plan architecture among all the invertebrate groups.
- Graptolites are planktic colonial organisms of Paleozoic age; they are among the most accurate fossils used in intercontinental correlation due to morphologic variability and high evolution rates.
- Conodonts are the most frequent debris from the cephalochordates and allied groups; a notochord is the axial skeleton of these organisms.
- Vertebrates have the axial skeleton consisting of the vertebral column; the skull situated in the anterior part of the body has the primary role of protecting the brain.
- Vertebrates were exclusively marine during the Early Paleozoic, but adapted to the freshwater and terrestrial environments in the latest Devonian times, possibly earlier.

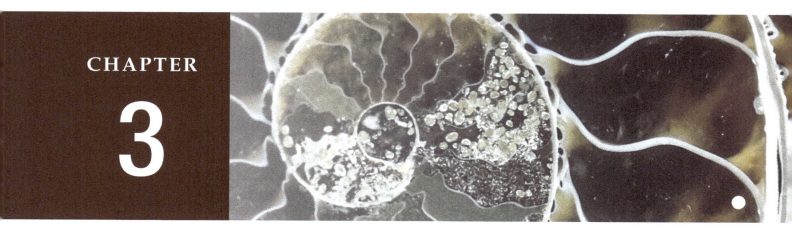

CHAPTER 3

FOSSILS AND THEIR APPLICATIONS

CONTENT

3.1 Biostratigraphy
3.2 Paleobiogeography
3.3 Paleoecology and Paleoenvironment Reconstructions
3.4 Fossil Record Applications in the Theory of Evolution Study
3.5 Fossil Uses in Economy

Chapter Conclusions

3.1 Biostratigraphy

The *stratigraphic range* of a certain fossil is defined as the interval between its lowermost and uppermost occurrences; each fossil species has its own stratigraphic range. The lowermost occurrence (LO) is also known as first appearance datum (FAD), last downhole occurrence (LDO), or evolutionary occurrence (EO); the uppermost occurrence is also referred to as last appearance datum (LAD), first downhole occurrence (FDO), or extinction (EX). Biostratigraphy is the branch of stratigraphy that uses fossil stratigraphic ranges and bioevents that define them (e.g., FAD and LAD, etc.) for dating and correlation.

Not every species can be used successfully in correlation; in order to be useful in and be therefore considered an *index species* or *marker*, a species must present six characteristics.

- Be *frequent* and even abundant in the fossil record.
- Have a *relatively short stratigraphic range*; therefore it is preferable to belong to a fossil group with a fast rate of evolution.
- Be *easily recognizable* in an assemblage; sharp morphologic features separating it from the ancestor and descendant species represent major criteria in selecting an index species.

- Be preferably a *facies independent* fossil; planktonic and nektonic groups provide the most accurate index fossils from this perspective.
- Present a *fast rate of geographical dispersal* and *colonization* of new geographical areas.
- Present a *wide geographical distribution*, in order to be suitable for regional and intercontinental correlation.

The fundamental unit in biostratigraphy is the *biozone*; biozones are often referred to as *zones*. A biozone is defined as the stratigraphic interval between two bioevents. There are several kinds of biozones that are used in biostratigraphy and of them the most frequently used are presented: assemblage zones, taxon range zones, and interval zones.

- *Assemblage zones* are defined in general by the overall fossil content; these are the oldest biozones used in biostratigraphy, and a large number of species is used in defining them (Figure 3.1). The *standard assemblage zones* (AZ) were used for the first time by Sir William Smith (1769–1839) in mapping England, Wales, and southern Scotland at the beginning of the nineteenth century; the basis for the definition and use of this kind of biozones is represented by the total fossil content of a layer, which makes it different from the fossil content of other layers, adjacent to it or not. This kind of biozone is frequently used even today, especially in sediment and rock dating for industry purposes. The *Oppel zone* (OZ) is another kind of assemblage zone, and was defined by the German paleontologist Carl Albert Oppel (1831–1865); in contrast with the standard assemblage zones, an Oppel zone has the lower and upper boundaries defined by first occurrences or last occurrences of two or more species; Oppel zones are not in the current use due to the many bioevents necessary in their definition.
- *Taxon range zones* (TRZ) are defined with the aid of one species, which is also the nominal or index species. A TRZ represents the stratigraphic interval between the FAD and LAD of the index species (Figure 3.1).
- *Interval zones* (Figure 3.1) are of two kinds, according of the number of species used in their definition: standard interval zones and partial taxon range zones. The *standard interval zones* (IZ) are defined with the aid of two species; there are four situations according to the type of bioevent nature (two FADs, two LADs, one FAD, and one LAD with the LAD situated above the FAD, and one FAD and one LAD with the LAD situated below the FAD). The index species in the cases with two FADs and LADs is considered that species with longer stratigraphic range; both species are considered markers in the cases with a combination of FAD and LAD. Three species are used in the definition of a *partial taxon range zone* (PTRZ); a partial taxon range zone represents the stratigraphic interval with continuous occurrences of one taxon, which is also the biozone index species, between two bioevents (FADs or LADs defined by other species as in the standard interval zones). The index species in the case of the PTRZ is considered the species with the longest stratigraphic range.

A biozone has a limited distribution in space, which largely corresponds to the extension of the index species. The stratigraphic range of the index species varies from sedimentary basin to sedimentary basin, and the FAD and LAD at various geographical locations depends of colonization and abandonment of certain basins and regions; therefore, the distinction between the *biozone*, which is defined by the evolutionary occurrence and extinction of a taxon, and *local biozone*, which is the result of the colonization and abandonment of a geographically restricted zone, should be made.

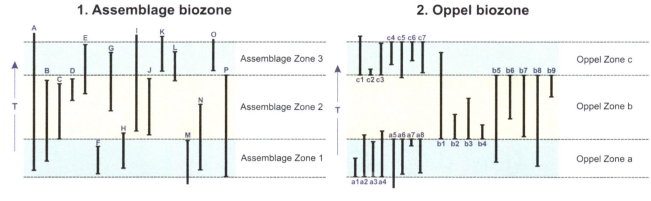

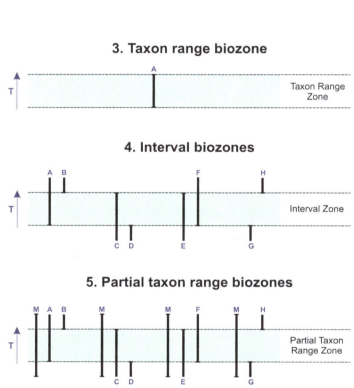

Figure 3.1 Main types of biozones used in biostratigraphy.

3.2 Paleobiogeography

Paleobiogeography represents the study of the organism distribution in space, as well as the changes in time of the distribution areal. A considerable amount of data on organism distribution in time was available to scientists at the beginnings of the nineteenth century; therefore, it became possible that biogeographical data provided significant evidence in support of the Theory of Evolution proposed by Darwin (1859).

Organism distribution in space and has two components: *biogeography* and *paleobiogeography*.

- *Biogeography* (*ecologic biogeography*) represents the study of the organism distribution as inferred from the modern climatic and ecologic settings; it is applied to the modern organisms and those from the recent past. *Provinces* and *realms* are large-scale subdivisions in biogeography; a lower scale subdivision in the case of plants is the *biome*.

- *Paleobiogeography* (*historical biogeography*) represents the study of the organism distribution that cannot be related to the modern climatic and ecologic factors; it is applied only to extinct organisms.

A common method in paleobiogeography is reconstructing a species distribution areal; this is realized by plotting the known occurrences on a paleogeographic map (Figure 3.2). These data represent the basis for reconstructing paleobiogeographic provinces and realms.

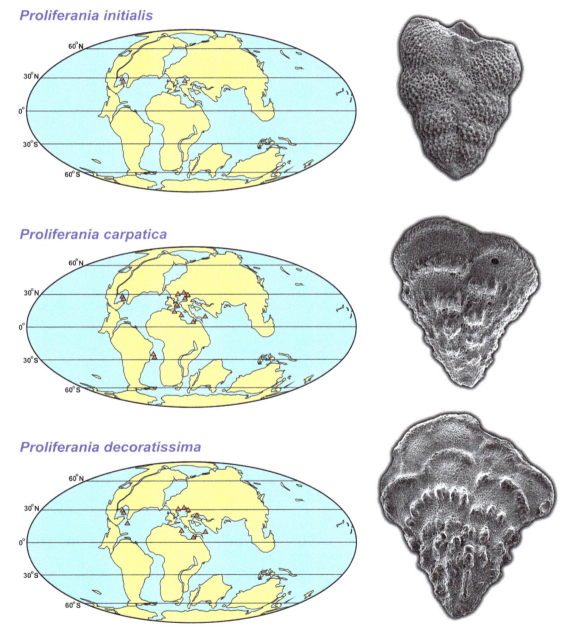

Figure 3.2 Fluctuation of the geographical distribution of the Late Cretaceous planktic foraminiferal lineage *Proliferania*. Note the rare occurrences in a narrow geographical region of the earliest species (*P. initialis*), maximum spread achieved with the evolution of sutural ridges (*P. carpathica*) and areal reduction with the evolution of the multichamber growth in the adult stage (*P. decoratissima*). After Georgescu (2010).

Glossopteris
Permian,
Australia

Figure 3.3 Seed fern *Glossopteris*. Specimen from the paleontological collections of the University of Calgary.

Species distribution played a paramount role in the reconstruction of the Earth geological history. The occurrences of the seed fern *Glossopteris* (Figure 3.3) in the Late Paleozoic rocks of the continents of the southern hemisphere were used by E. Suess (1831–1914) to infer the existence of a supercontinent, he named Gondwanaland. Additional data on the reptiles *Mesosaurus* (South America and southern Africa), *Cynognathus* (South America and Africa), and *Lystrosaurus* (Africa, India, and Antarctica) further helped A. Wegener (1880–1830) to demonstrate the ocean-floor spreading and the existence of Gondwana.

3.3 Paleoecology and Paleoenvironment Reconstructions

Paleoecology fundamentally represents the ecology of the fossil species and communities, and largely represents the "ancient" equivalent of ecology that studies the modern ecosystems. There are two components within paleoecology: *paleoautecology* and *paleosynecology*.

- *Paleoautecology* studies the relationships between a fossil species or a community of fossil species and the biotic environment in which it lived and evolved.
- *Paleosynecology* represents the study of the relationships between a certain species or fossil community and the abiotic environment in which these entities existed.

Both paleoautecology and paleosynecology data present importance in a correct reconstruction of a fossil species and/or community paleoecological relationships. The paleoecological data are useful in reconstructing the sedimentary paleoenvironments; *paleoenvironmental reconstructions* are interdisciplinary studies, which take in consideration the data from a variety of geological disciplines: paleontology, sedimentology, tectonics and structural geology, geochemistry, and paleoclimatology, etc.

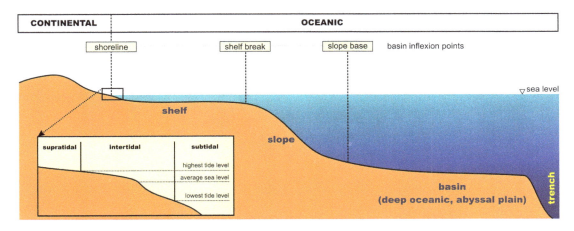

Figure 3.4 Classification of the sedimentary domains of an oceanic basin.

Paleoenvironmental reconstructions are extensively used in the exploration for hydrocarbons. An example of paleoenvironmental reconstruction is represented by the paleobathymetry estimations; this is done by recognizing in the fossil record key species or assemblages that indicate sedimentation in certain conditions, such as *fresh water*, *brackish water*, and *marine environments*.

- *Fresh water* species are recorded in waters with low salinity values, which occur in lake, ponds, and river conditions.
- *Brackish water* environments indicate in general sedimentation in the proximity of the boundary between the marine and continental realms.
- *Marine environments* are further subdivided according to the water depth; the basic separation is into shelf (0 to ~200 m, occasionally down to a depth of about 500 m), slope (~200 to ~500 m, occasionally to a depth of about 2500 m), and basin settings that extend to the deepest portions of the deep seas and oceans (Figure 3.4). The environment separation in the seas and oceanic basins into shelf, slope, and deep basin is according to the three inflexion points of an oceanic basin (shoreline, shelf break, and slope base).

Paleoenvironment recognition is often complicated the occurrences of the assemblages with evident morphologic similarities in distinct paleoenvironmental settings. For example, a foraminiferal assemblage consisting only of agglutinant species can occur in estuarine shallow waters (water depth <30 m) (Figure 3.5) and deep oceanic basin (water depth >3000 m) (Figure 3.6). In the latter case the calcareous tests are not fossilized due to extensive dissolution that occurs under the lysocline.

3.4 Fossil Record Applications in the Theory of Evolution Study

One of the major applications of paleontology is the study of the Theory of Evolution; this application should be considered in a broader context, paleontological data complementing those from other sciences such as biology and genetics. The use of paleontological data in understanding the geological history of life on Earth began long before the release of the Theory of Evolution. For example, based on the data from the fossil record Georges Leopold Cuvier (1762–1832) and Alexandre Brogniart (1770–1847) demonstrated that species can become extinct; a similar conclusion was

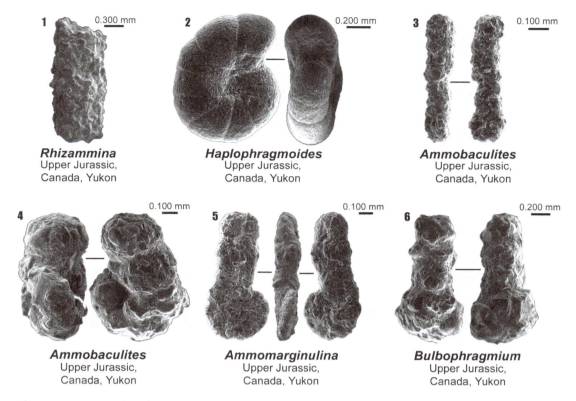

Figure 3.5 Examples of estuarine shallow water agglutinant foraminifera of Late Jurassic age from northern Canada. All specimens illustrated by Georgescu and Braun (2013).

drawn by Sir William Smith (1769–1839) when realizing the geological map of England, Wales, and southern Scotland. Such observations showed the fact that the species known in the fossil record are not unchangeable entities that originated at the time of religious creation as considered in the past when creationism was a standard in science. One of the earlier ideas of evolution in the living world was proposed and theoretically developed by Jean Baptiste Lamarck (1744–1829), who considered that species can evolve from each other as the result of an *inner want*; this idea was not accepted in the scientific community mainly because the existence of such an inner want could not be demonstrated.

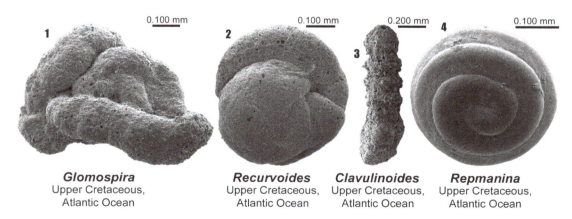

Figure 3.6 Examples of deep-water agglutinant foraminiferal taxa from sediments of Late Cretaceous age of South Atlantic Ocean (Deep Sea Drilling Project, Site 511, Falkland Plateau).

The Theory of Evolution was developed by Charles Robert Darwin (1809–1882) in his book titled *The Origin of Species by Means of Natural Selection or the Preservation of Favoured Races in the Struggle for Life*; the first edition of this book was published in 1859. Therefore, this year is considered the date of birth of the Theory of Evolution. Through independent study Alfred Russel Wallace (1823–1913) provided the hypothesis of the Theory of Evolution in an article published in 1855; notably, Wallace considered at a lesser extent than Darwin the arguments given by the paleontological data in developing and promoting his discovery. Darwin (1859) dedicated two chapters to the paleontological data and their support of the Theory of Evolution, namely chapter IX (*On the Imperfection of the Geological Record*) and chapter X (*On the Geological Succession of Organic Beings*). In his work Darwin emphasized both the importance of using the paleontological data for the study of evolution and the difficulties in understanding this process at the beginnings of the second half of the nineteenth century. Among the major problems that some authors called later *Darwin's dilemmas* and existed at the time when the Theory of Evolution was released are: absence of a mechanism to explain the intraspecific variability, lack of knowledge on fossil organisms prior to the Proterozoic/Phanerozoic boundary, absence of an evolutionary classification methodology, etc. Despite such dilemmas, as new data were acquired in the next decades it became evident that only paleontology and the Theory of Evolution can provide a rational explanation for the fossil record.

The continuous development of ideas in the field of paleontology significantly contributed to the general understanding of the Theory of Evolution; data influx and conceptual development led to the new stage in the development of the Theory of Evolution. This new stage is known as the *Modern Synthesis* and was developed in the first half of the twentieth century. Modern Synthesis is the scientific current in which the study of life evolution was separated between three branches of natural sciences: biology, genetics, and paleontology; this process fundamentally acknowledged the differences in the data nature used by the three sciences. Three personalities dominated biology, genetics, and paleontology during the Modern Synthesis: Ernst W. Mayr (1904–2005), Theodosius G. Dobzhansky (1900–1975), and George G. Simpson (1902–1984), respectively. Modern Synthesis brought a significant increase in understanding the patterns in the life evolution on Earth, *tempo of evolution*, etc. In addition, the taxonomic studies revealed the necessity to define the concept of species applied in paleontology as "… a lineage (an ancestral-descendant sequence of populations) evolving separately from others and with its own unitary evolutionary role and tendencies" (Simpson, 1963, p. 153). One significant advance in the times of Modern Synthesis is the separation between the evolution at the scale of large groupings of species (*macroevolution*), and evolution at species and below the species level (*microevolution*).

3.5 Fossil Uses in Economy

Fossils applications in economy began in the early nineteenth century when Sir William Smith used them in correlating layers of sedimentary rocks cropping out in distant sections; this correlation method was developed during the mapping of England, Wales, and southern Scotland. Correlation with the aid of fossils (*biocorrelation*) is based on the fact that the existence in space and time of one fossil species is unique and once a species became extinct it will never reappear again in the fossil record.

Using fossils for *sedimentary rock dating* was a gradually improving process. At first the invertebrate fossil were used for such a purpose and ammonites, bivalves, trilobites, and graptolites were amongst the most frequently used groups; this setting lasted for the most part of the nineteenth century. The challenge induced by the public illumination and invention of the engine with internal combustion determined a significant increase of the demand for hydrocarbons in the last decades of the nineteenth century; the search for liquid and gaseous hydrocarbons was made initially by digging and subsequently drilling wells. Macrofossils were of limited help in dating the sediments drilled because most of the large-sized fossils were crushed by the drilling bit. As a result of this challenge, microfossils and especially foraminifera started to be were widely used, an application that was pioneered in Poland by József Grzybowski (1869–1922). The peak in oil production in the 1930s and 1940s is largely due to the foraminiferal studies that resulted in a high accuracy in sediment dating and consequently identification with precision of the subsurface structures capable to host oil and natural gas. Sediment dating by means of microfossils was refined in the second half of the twentieth century when other groups such as radiolarians, green algae, conodonts, ostracods, diatoms, silicoflagellates, nannofossil, etc. started to be used on large scale.

A new challenge for the use of microfossils in industry was induced by the development of sequence stratigraphy in the late 1970s, which was rapidly adopted as standard method in exploration for hydrocarbons. *Sequence stratigraphy* considers that the distribution of the various layers of sedimentary rocks is controlled by sea-level changes; the sea-level changes were produced in the geological past by a combination of tectonic and climatic factors. Fossils and especially microfossils started to be used for a double purpose in order to understand the layer formation and therefore, reservoir architecture in the context of sequence stratigraphy: for *sediment dating* and in *paleobathymetry estimations*. The quantitative use of microfossils in recognizing the transgressive-regressive cycles was pioneered by Israelsky (1949).

Paleoecological data play a major role in industry their applications followed the release of the concept of *depositional system* in the late 1950s. In a classical study Stehli and Creath (1964) followed the planktic/benthic foraminifer ratio to recognize the ancient topographical features of the sea-floor, paleo-current directions, structural sea-floor features and their evolution in geological time and predict the existence in the geological past of layers that were removed through erosion.

Recognizing the regularities in the distribution patterns of the fossils at the scale of the complete evolution of a basin represented the next step in fossil use for hydrocarbon exploration; this type study was realized for the first time in the Western Black Sea (Georgescu, 1997, 2003) and demonstrated that fossil distribution can be used successfully in recognizing basin evolution phases and major tectonic events in the basin history. Such studies are based on weighted samples collected from sections and wells distributed over the entire surface of a basin. Three kinds of fossil distribution can be recognized (Figure 3.7).

- *1st type distribution* occurs after the beginning of the basin evolution, namely during and shortly after rifting; fossils are rare (maximum a few specimens/unit of sample) and barren samples are frequent. This distribution type occurs during the flysch sedimentation phase. Microfossil assemblages are dominated by primitive species; foraminiferal assemblages consist almost entirely of agglutinant species and radiolarians are frequent occurrences; the 1st type distribution is characteristic to the deepest sedimentation in the basin evolution. Fossil *dissolution* is the dominant process in the 1st type distribution.

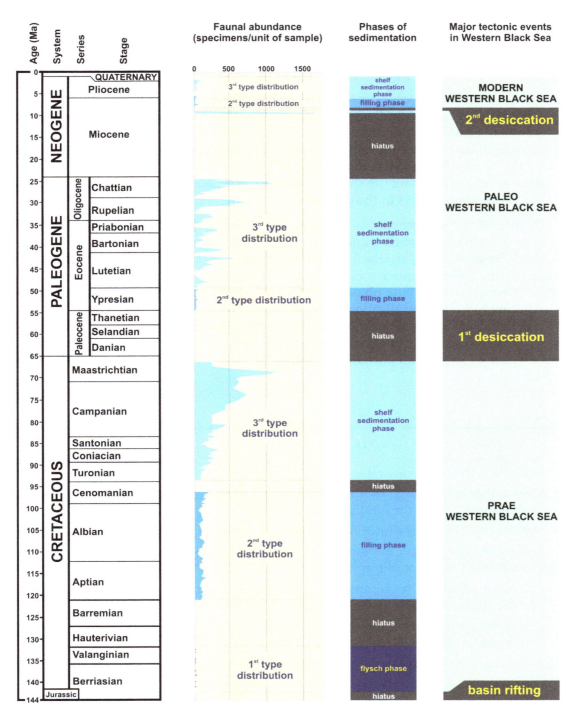

Figure 3.7 Example of phases and major events recognized in a basin evolution with the aid of microfossils; case of the Western Black Sea, based on Georgescu (2003).

- *2nd type distribution* occurs during the filling phase; deep sea turbidite sediments accumulated at the slope base are frequent during this phase, which marks the beginning of the oceanic basin filling. Although accumulated in the deep oceanic environments the sediments of the filling phase are shallower than those formed during the flysch phase. Basin infill during the filling phase is due to the massive transport of sediments from the adjacent continental areas. Microfossil assemblages are dominated by the reworked organic debris; in situ microfossils (e.g., foraminifers, radiolarians, diatoms, ostracods, etc.) occur only in the sediments formed during the periods with reduced or absent terrigenous material input. *Reworking* and *transport* are the two processes that dominate in the 2nd type distribution.

- *3rd type distribution* occurs in the shelf sedimentation phase, which corresponds to the terminal basin evolution phase when the basin is almost completely filled with sediments. Sediment accumulation during the shelf sedimentation phase happens is the shallowest during basin evolution and fossil assemblages are characterized by the highest abundance and diversity. The vast majority of the fossils forming the assemblages during the shelf sedimentation phase are in situ; reworked fossils are minor components in the paleontological assemblages.

By following the succession and characteristics of the three types of fossil distribution it is possible to recognize the major events in the basin evolution. In addition, it is possible to predict the accumulation of the source, reservoir and seal rocks; this is based on the fact that source rocks dominate in the flysch phase, reservoir rocks accumulate mainly in the filling phase, and seal rocks in the shelf sedimentation phase.

Another industrial application of the fossils in the estimation of the rock depth of burial; this application is based on the fossil property to change their colour with the temperature increase during the burial process. Organic-walled algae and conodonts, which also contain small amounts of organic matter, are particularly useful in such applications.

CHAPTER CONCLUSIONS

- Biozone is the fundamental unit in biostratigraphy.
- The stratigraphic range of a fossil is defined between its lowermost and uppermost occurrences in the stratigraphical record.
- Species used in biostratigraphy are known as index species or marker species.
- Assemblage zones, taxon range zones, and interval zones are the most used types of units used in biostratigraphy.
- Paleobiogeography represents the study of the organism distribution in space, and the changes in time of the distribution areals.
- Paleoecology represents the study of the fossil species ecology; it has two components: paleoautecology and paleosynecology.
- Fossil record is of paramount importance in the study of the Theory of Evolution.
- Fossils are extensively used in economy in sediment dating, paleoecological reconstructions, relative sea-level fluctuation reconstructions, and recognition of the sedimentary basin tectonic evolution phases.

CHAPTER 4

FOSSIL RECORD AND LIFE EVOLUTION

CONTENT

4.1 From Biological Population to Paleontological Assemblage
4.2 Intraspecific Morphologic Variability
4.3 Types of Specimens in Paleontology
4.4 Species in Paleontology
4.5 Speciation, Species Evolution, and Extinction
4.6 Principles and Methods of Species Classification
4.7 Macroevolution

Chapter Conclusions

4.1 From Biological Population to Paleontological Assemblage

Studying the life evolution on Earth implies not only a significant amount of knowledge on the living organisms and the vestiges of the ancient life forms in the fossil record, but also a correlation between the two components; by understanding the processes through which the organisms become fossils we can understand how much of the totality of the once living organisms had the potential to become fossils, and how much of the data pool represented by fossils is actually used in the paleontological studies.

Fossilization is only a part of the set of processes that shapes the fossil record. There are geological processes at the Earth surface or within the crust and mantle that strongly modify the fossil record. For example, small sedimentary basins with an area of about several tens of square kilometres are formed in transform settings, where lithospheric plates move along each other and there is no apparent subduction between them. These basins have a short lifespan, which ranges from 1 to 10 million years; the destruction of a small-sized such basin in the case of re-activated subduction will result in its complete melting (fundament and sedimentary cover rocks) in the asthenosphere. All the fossils formed during the basin life span are obliterated from the fossil record in such a case. Moreover, in the case of the *endemic species*, which had the distribution area restricted to that particular sedimentary basin, they will be completely obliterated from the fossil record. The process of weathering of the sedimentary and metamorphic rocks that contain fossils is another source of fossil record bias; as the layers or rocks are weathered and eroded the portion of the fossil record they host is mostly completely destroyed and cannot be reconstructed.

Fossil classification and fossil record study require additional knowledge on the organism transformations in the course of the fossilization process. Biologists who study the living organisms have usually a large number of specimens available, and often abundant populations; frequently it is possible to study the population succession over a narrow time interval, which ranges from several days to tens of years. Therefore it is possible, and often is a requirement in certain biological specialities to analyze the *biological population* or an array of biological populations that form the species. Such level of detail is not available to a paleontologist who studies only a small part of the once living organisms. The biases in paleontological studies are induced by the nature of the fossil record.

- *Fossilization bias*. The fossilization rate is frequently of about 1:1000 to 1:1,000,000, sometimes even smaller. For example, if sedimentation in the oceanic realm happens below the lysocline, the calcitic, and aragonite tests of planktic organisms will be completely dissolved and none of them or rare fragmentary tests will remain in the fossil record.
- *Areal bias* is given by the sampling technique; although a layer containing fossils can be spread over several square meters to millions of square kilometres, only a small portion of it is studied. For example the species recorded in a layer intersected by a well in offshore conditions can be studied only by the specimens occurring in the samples obtained from that particular well with a section area of a few square centimeters; additional specimens can be collected if new wells are drilled, but the drilling decisions are taken according to economic factors.
- *Sampling bias* may occur when only a portion of a layer is sampled; continuous sampling in which the samples are collected without leaving intervals between them prevents the occurrence of this type of bias.

These limitations are part of paleontological studies. As a result of them, paleontologists do not work with populations, but with paleontological assemblages. A *paleontological assemblage* is a biased succession of biological populations. Paleontologists use sometimes term "population" in describing a group of individuals collected from the same site (outcrop or well) from one stratigraphical level; technically this is incorrect and the correct term should be that of paleontological assemblage.

4.2 Intraspecific Morphologic Variability

Recognizing species in the fossil record is a major task for every paleontologist involved in the study of any group of fossil organisms. This challenge is primarily generated by the incompleteness of the fossil record and the lack of biological and genetic data, which are necessary to assess the potential interbreeding between the members of a fossil assemblage. The methodology of species recognition frequently includes careful systemic observations on the morphology of specimens collected from stratigraphically

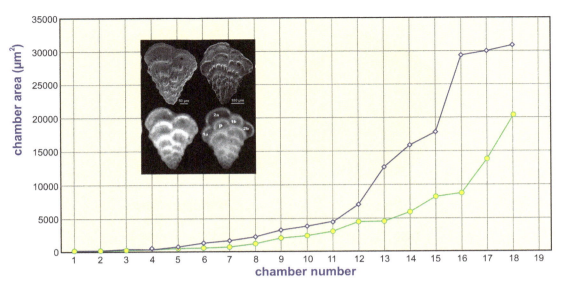

Figure 4.1 Ontogenetic trajectories of two planktic foraminifera realized by plotting the chamber surface area. Abbreviations: P-progressive chamber, 1- and 2-chamber sets of the multichamber growth stage.

ordered samples. Intraspecific variability should always be taken in consideration in the course of fossil species identification. There are three kinds of differences between the individuals of one species: ontogenetic, genetic, and nongenetic.

Ontogenetic variability is the result of organism growth during its lifetime. There are frequent cases where there are significant morphologic differences between the juvenile, adult, and gerontic specimens of the same species. The correct identification of these growth stages is of paramount importance in species identification and prevents the artificial proliferation of species. There are two types of growth that are defined according to the organism shape change during the ontogenetic development: isometric and anisometric. The two growth types can be recognized by plotting two parameters pertaining of the organism as a whole or a particular morphologic structure. The series of plots resulted from measuring the specimens of a paleontological assemblage can be used to define the *ontogenetic trajectory* (Figure 4.1). *Isometric growth* occurs in the species in which the shape remains unchanged during the growth process; in this case the ontogenetic trajectory is a straight line (Figure 4.2). *Anisometric growth* occurs in the species in which there is a change in

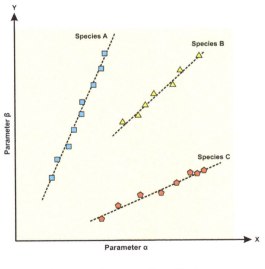
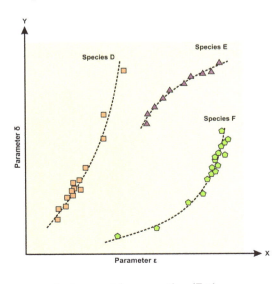

Isometric growth α/β=k   Anisometric growth ε/δ≠k

Figure 4.2 Diagram presenting idealized cases of isometric and anisometric growth,

Dosinia
Miocene,
USA, Maryland

Heliophyllum
Devonian,
USA, Montana

Figure 4.3 Example of growth through accretion in the case of a bivalve (left) and a rugose coral (right). Both specimens from the paleontological collections of the University of Calgary.

shape during the ontogeny, and therefore the ontogenetic trajectory is represented by a curve line (Figure 4.2).

Growth strategy differs between the various groups of organisms. Five kinds of growth strategies are known: accretion, addition, molting, modification, and mixed growth.

- *Accretion* is characterized by continuity in the development of skeletal parts. This growth strategy occurs in the gastropods, bivalves, brachiopods, etc., and can be easily recognized by the development of growth lines over the shell surface (Figure 4.3).
- *Addition* occurs in general in the organisms in which the skeletal parts consist of a larger number of components; addition of new such smaller skeletal components is necessary with the soft body growth. This growth is typically developed in the all the echinoderm groups and occurs occasionally in certain arthropod species (Figure 4.4).
- *Molting* occurs in the groups of organisms in which the exoskeleton is rigid and cannot accommodate the rapidly growing soft body because new parts are not added; therefore, the organism must renounce periodically at the constraining exoskeleton and produce another one that is larger. Molting occurs frequently amongst arthropods (e.g., trilobites, insects) (Figure 4.5).

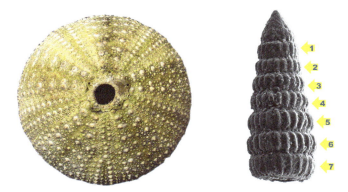

Figure 4.4 Example of growth through addition in the case of a regular echinoid (left) and a radiolarian (right). Specimen to the left is from the paleontological collections of the University of Calgary.

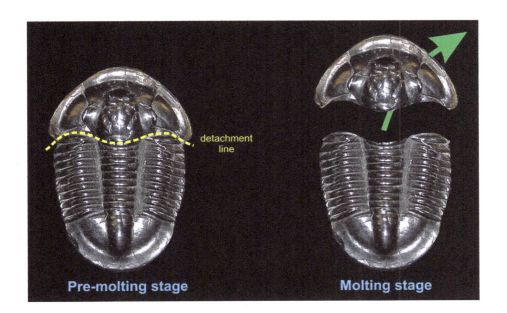
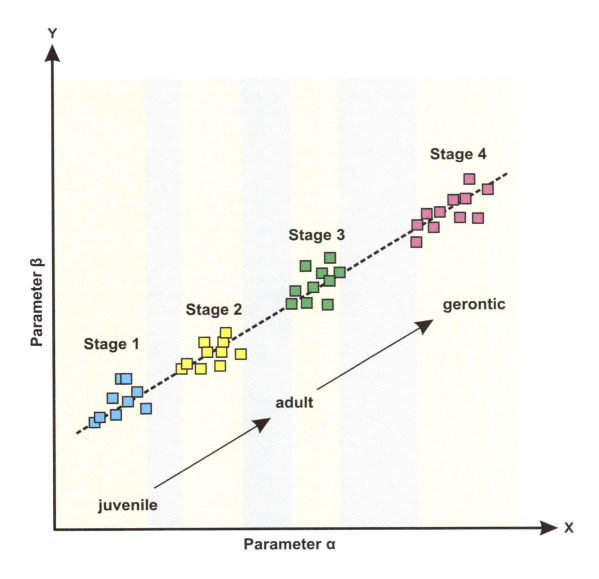

Figure 4.5 Growth through molting. Above - idealized and artificial representation of the trilobite carapace separation during molting. Below - idealized diagram showing the plot clustering in the case of growth through molting.

Nautilus
Recent,
Pacific Ocean

Figure 4.6 Example of mixed growth in the case of cephalopod genus *Nautilus*. Left-external view showing growth through accretion. Right-internal view showing growth by addition.

- *Modification* is frequent among the vertebrates, where the bones change their shape with the size increase.
- *Mixed growth strategy* occurs through the combination of accretion and addition. It is typically developed amongst the nautiloid and ammonoid cephalopods where the soft body is situated in the last-formed chamber; growth through accretion occurs at the last-formed chamber, whereas the addition strategy is documented by the formation of septae (Figure 4.6).

Genetic variability is a direct reflection of the species genetic structure. Genetic differences can be continuous or discontinuous.

- *Continuous genetic variability* refers to the morphologic variability exhibited by each feature of a species and it can be presented quantitatively through measurement data. It is evident when we study a large number of individuals of the same species that practically each feature has a more or less expressed variability, which is very important in species description. In order to assess properly the taxonomic significance for each morphologic feature of a species its variability must be assessed in each sample and successive samples. The continuous morphologic variability occurs in all the populations of the same species and cannot be directly related to certain climatic conditions, biotic stress, and geological events. Specimen sizes in a species, dimensions of certain morphologic features, ratio between two measurable features and color are examples of continuous morphologic variability (Figure 4.7).
- *Discontinuous genetic variability* is less frequent than the continuous one and is common especially in the species with differentiated sexes (*sexual dimorphism*). Male and female individuals are easily recognizable in arthropod and vertebrate species; morphologic differences are frequently evident and therefore, there is a distinct gap between the two morphologies. Sexual dimorphism is more difficult to demonstrate in the case of fossil species. For example, some ammonite species of the Jurassic and Cretaceous present larger and smaller shells with significant resemblances in the coiling mode and ornamentation, which occur in large numbers at the same stratigraphic levels; although in the absence of the soft body parts it is hard to demonstrate which are males and females, respectively, such

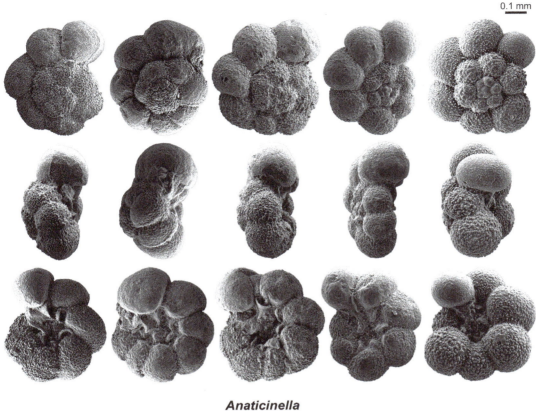

Figure 4.7 Example of continuous genetic variability in the case of Cretaceous planktic foraminifer *Anaticinella*. Upper row-spiral views, middle row-edge views, lower row-umbilical views.

differences are considered the result of sexual dimorphism. The discontinuous genetic variability resulted in the course of reproduction process is frequent in certain protistan groups. This situation is frequent among the benthic foraminifera that present an alternance between generations with sexual and asexual reproduction, which result in the formation of smaller tests with larger proloculus (macrospheres) and larger tests with smaller proloculus (microspheres), respectively. Another example of discontinuous genetic variability is the coiling mode among the planktic foraminifera. The tests can be coiled either to the left (sinistral) or to the right (dextral) and no transitions between the two types do exist (Figure 4.8).

Nongenetic variability is primarily induced by ecologic factors. Suprapopulation in the case of brachiopod clusters leads often to the occurrence of deformed shells due to the small space left at one specimen disposal; this is also the case of some bivalve species (Figure 4.9). The effects of suprapopulation on the foraminiferal tests are evident in specimen size: small-sized specimens are frequent in abundant assemblages due to the small amounts of available nutrients and large-sized specimens dominate the populations consisting of a small number of individuals due to the abundant nutrients. Nongenetic variability is frequent in the benthic foraminifera with agglutinated test; although the species included in this group are considered in general selective in the size of agglutinated particles from the surrounding environment, the

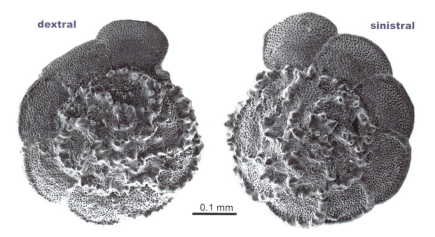

Figure 4.8 Example of discontinuous morphological variability in the case of the Cretaceous planktic foraminifer *Pseudothalmanninella*.

Figure 4.9 Example of suprapopulation in the case of one bivalve species from Shark Bay (Eastern Indian Ocean, offshore Australia).

specimens living on a sandy substratum will have a coarser appearance when compared to those living on a finer one due to the coarser material they use in building their test (Figure 4.10).

4.3. Types of Specimens in Paleontology

Species definition is of paramount importance in all paleontological studies in which such units are used. As a result a system of specimen naming was developed in order to assure a standard referencing system. This system is based on the reference to certain type specimens that are either selected by a species author or subsequently designated by other specialists who reviewed the species. Today we can assign each specimen of a certain species to a certain specimen type. The different specimen types are not hierarchically arranged. Their definition is standardized through the

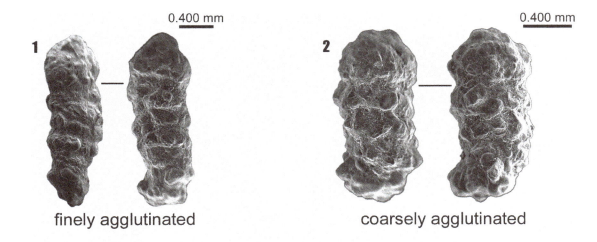

Figure 4.10 Example of non-genetic variability in the case of one agglutinant benthic foraminiferal species of Late Jurassic age from northern Canada. Specimens illustrated by Georgescu and Braun (2013).

International Code of Zoological Nomenclature (ICZN) and International Code of Botanical Nomenclature (ICBN). All the terminology below refers to the specimens of the same species.

- *Holotype* is specimen designated the species name-bearer by the species author; the author is the specialist/specialists who proposed a formal name for the species. The holotype is unique and designated from the original population in biological sciences or paleontological assemblages in the paleontological sciences; only the author can designate a holotype. The common practice is to select as holotype the best preserved specimen in the population or assemblage; some authors prefer to select the holotype amongst the mature rather than gerontic specimens (Figure 4.11). Holotype does not necessarily present the totality of features of all specimens included in the species.

- *Paratypes* are the remaining specimens from the biological population or paleontological assemblage after the holotype was selected. The number of paratypes is variable from species to species. Paratypes cannot encompass completely the species variability, but they provide a pool of specimens for the holotype replacement in case the latter is lost.

- *Neotype* is the specimen selected by a specialist in the case the holotype was lost or destroyed. Holotype loss/destruction must be very well documented and it is the species reviser duty to provide compelling evidence in order to demonstrate the species name-bearer loss/destruction. Neotype must present similar morphologic features to those of the lost or destroyed holotype and this requirement can be achieved by comparing it to the original and subsequent illustrations of the holotype. An ideal case is when the neotype is selected from amongst the paratypes. In case paratypes are not available it is preferable to having the neotype selected from the same geographic area and from sediments of the same age from which the holotype was originally collected and designated. In case the neotype is lost, a new neotype can be selected.

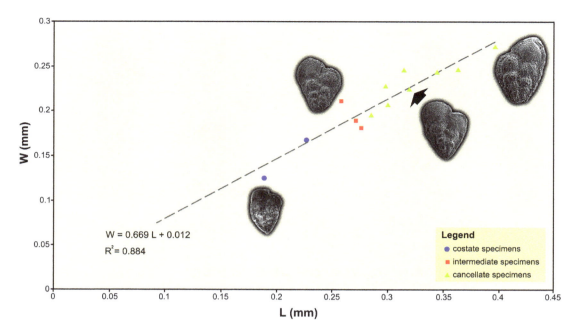

Figure 4.11 Example of holotype selection from amongst the mature individuals of the Late Cretaceous planktic foraminiferal species *Hendersonites pacificus*.

- *Lectotype* is selected by a species reviser after a species was formally described and when the author did not select a holotype. Such situations often occur in the case of older species, for example those published in the nineteenth century and the beginning of the twentieth century, prior to the publication of the first edition of the ICZN and/or ICBN. The lectotype can be selected by the species reviser either from the figured specimens in the original publication or from the original material in case it is conserved in a collection.
- *Paralectotypes* are the remaining specimens after the lectotype selection.
- *Topotype* is a specimen from the same location (well or outcrop) as the original type specimens (holotype and paratypes) usually by another collector; frequently topotypes are collected a long time after the type specimens. Specimens from a nearby outcrop or well are commonly referred to as *nearly topotypes*.
- *Hypotype* is a specimen from any location (well or outcrop) other than the type location, and from sediments or sedimentary rocks of the same or different age.

The definition of type specimens involves additional concepts, namely those of type locality and type level (Figure 4.12).

- *Type locality* is the location (outcrop or well) from which the holotype and paratypes are collected. If a specimen is collected from a different locality, then it is considered a hypotype. A specimen collected from the same locality, but after the species formal description is a topotype. In the case that the original outcrop was destroyed (e.g., covered by the waters of a dam, etc.) then a specimen collected from a nearby locality can be referred to as "nearly topotype," but this is an informal expression.
- *Type level* is the layer from which the holotype was collected, must be mentioned in the species formal description. Specimens collected from different layers from the type locality should be considered topotypes.

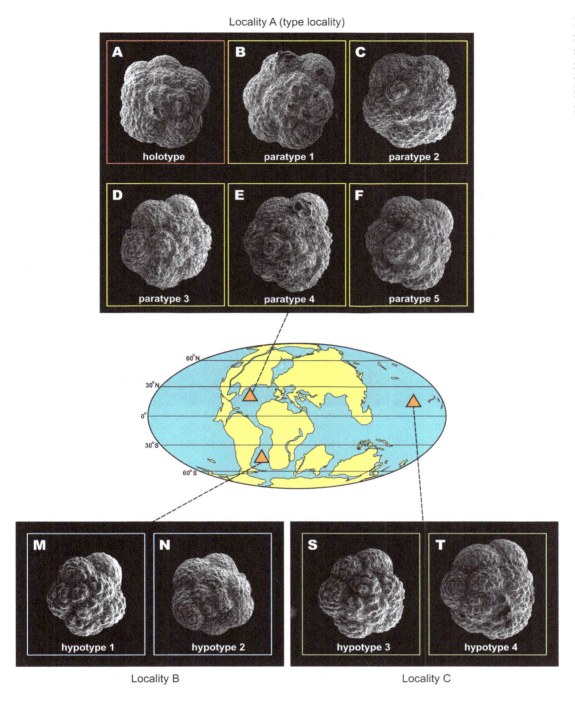

Figure 4.12 Idealized representation of the some specimen types, namely holotype, paratypes and hypotypes in the case of an alleged new species.

An example of significant complexity is herein presented; it is the case of the species originally described as *Textilaria americana* by Ehrenberg (1843) and figured later by Ehrenberg (1854) (Figure 4.13). This author did not designate a holotype either in 1843 at the species formal description or in 1854 when one specimen was illustrated. The specimen can be designated lectotype; it is in repository in the Ehrenberg Collection from the Naturkundemuseum (Berlin). The specimen is caught in Canada balsam and for this reason its ornamentation and test wall ultrastructure cannot be properly studied. The rediscovery of the original samples creates the possibility of defining a neotype from the detachable specimens, in order to have a reference

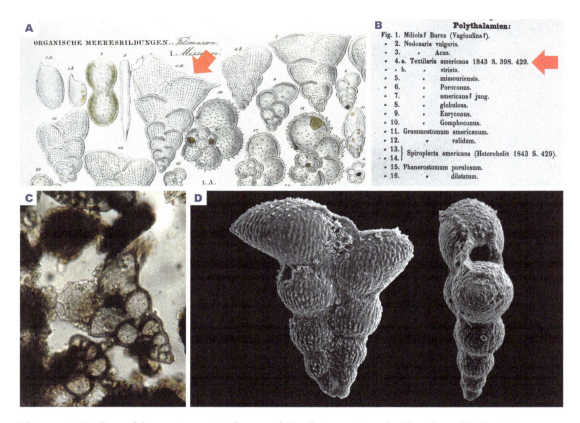

Figure 4.13 Case of the species originally named *Textilaria americana* by Ehrenberg (1843). A–B: Original illustration and caption by Ehrenberg (1854). C: Original specimen preserved in the Ehrenberg Collection. D: Possible neotype recovered from the same bulk sample.

specimen with these features that are of paramount importance in the group modern taxonomic framework.

Species documentation standardization raised a series of problems in the current practice of species identification and naming; two of them are briefly presented.

- A significant number of specialists use to illustrate only specimens that have close morphologic resemblances with the holotype, neotype and lectotype, upon case; this practice is often called the "*holotype syndrome*." The detrimental effect is that less information is added to improve our knowledge on species variability.

- Overemphasizing the role of the holotype, neotype, and lectotype induced the belief that a species is centred on these type specimens; therefore, a lot of specialists understand the paleontological species by reference to one specimen. Such practice that reminds of Plato's ideas, which is an idealistic philosophy abandoned thousands of years ago, should be avoided. The main argument against this practice is that the holotype is selected with the species formal description and cannot be changed afterward to accommodate the newly acquired knowledge of the species morphology.

Species are complex entities consisting of individuals clustered at various subspecific levels into populations, subspecies, etc.; the complexity is further expanded when we take in consideration the fact that every species has a distinct evolutionary history and trends that are recognizable in the morphologic features. A distinct solution to encompass a more significant portion of the morphologic variability of the species in space and time is to define a *type series* of individuals that van be improved by adding or extracting specimens from it, as our knowledge advances.

4.4 Species in Paleontology

Species are considered today the fundamental units in the classification of the life forms on Earth; this means that each living or fossil organism on our planet belongs to a certain species. The concept of species received various interpretations and was differently defined by scientists across the biological and paleontological sciences; the definition of the species concept is crucial in developing well-documented and precise fundamental and applied studies, and for this reason it represents today one of the major topics of discussion and scientific debate amongst taxonomists.

In *Systema Naturae* C. Linnaeus grouped specimens into species according to the degree of morphologic resemblance, but he did not provide a definition for the species concept. Later, the term of *Linnaean species* was coined to recognize the species concept defined according to general morphologic resemblance. Linnaean species were considered originally unchangeable entities of divine origin. The French naturalist Jean-Baptiste Lamarck (1744–1829) questioned the existence of Linnaean species; this point of view was driven by detailed observations on the intraspecific variability. Lamarck's interpretation got prominence in the first half on the nineteenth century, when species started to be considered heterogeneous units. Although the existence of varieties within one species was noted by C. Linnaeus, it was Charles Darwin (1809–1882) who proposed the *subspecies level* as a taxonomic level below that of species and therefore, formalizing the species heterogeneity.

The concept of *morphospecies*, which is widely used by paleontologists today, is derived from that of Linnaean species. According to this concept a species is an entity defined by its morphologic characteristics, which are distinctive and can be used to separate it from other units of similar rank. Modern and fossil species morphologic features are included and can be plotted within the *morphospace* (Figure 4.14). A morphospecies is often considered the synonym of the *typological species*. The centre of

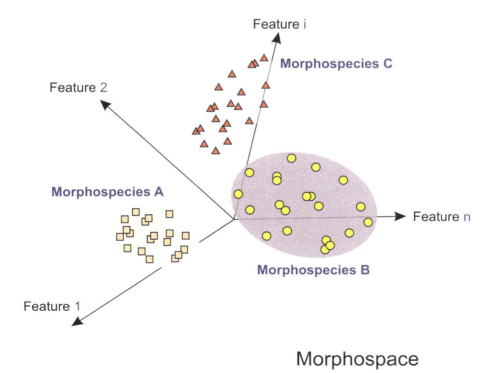

Figure 4.14 Idealized representation of three morphospecies within the morphospace.

weight in defining a typological species is put on the type specimens, which are frequently the holotype and paratypes; some researchers today even consider that a typological species is centred on the holotype. On the overall, the concepts of morphospecies and typological species should be considered the direct descendants of the Linnaean species concept.

A change at paradigm level in thinking about the history of life forms on Earth was when Darwin (1859) released the Theory of Evolution. According to this theory, species evolved from each other and the living world on Earth was and is in a permanent process of evolution. One of the most important achievements of Darwin was the transition from the static to dynamic understanding of life on Earth. The Theory of Evolution harmonized for the first time our understanding of the fossil and rock record on Earth by demonstrating that the species and higher taxonomic categories undergo a long process of slow transformation, which parallels the changes at the Earth surface as demonstrated by James Hutton (1726–1797).

The release of the Theory of Evolution triggered a new series of studies both in biological and paleontological sciences. Amongst the most significant developments in the next decades after the publication of *The Origin of Species* was the beginning of genetics, considerable developments in embryology and biological populations, etc. As a result, the Theory of Evolution as formulated by Darwin needed to be upgraded to a new level in order to incorporate the new achievements across the sciences dealing with the representatives of the living world.

The re-evaluation of the Theory of Evolution begun in the first half of the twentieth century and is known as the *Modern Synthesis*. During the times of the Modern Synthesis we witness a major split between biology, genetics, and paleontology, due to the accumulation of a vast amount of data in each of these components of natural sciences. This is apparent in the definition of three distinct concepts of species, namely the concept of biological species, genetical species, and paleontological species. Each of them represented the beginnings of a distinct stream in defining the species concept for that particular discipline.

- Probably the best known concept of biological species was given by Ernst W. Mayr (1904–2005) in the book *Principles of Zoological Classification*, published in 1969: "Species are groups of interbreeding natural populations that are reproductively isolated from other such groups."
- The original *concept of genetic species* was provided by Theodosius G. Dobzhansky (1900–1975) in his book *Genetics and the Origin of Species*, published in 1937. Today we consider a genetic species as an array of natural populations that are genetically compatible, and are genetically isolated from other units of similar rank.
- The concept of paleontological species was given by George G. Simpson (1902–1984) in *Principles of Animal Taxonomy* that was published in 1961: "… a lineage (an ancestral–descendant sequence of populations) evolving separately from others, and with its own unitary evolutionary role and tendencies."

There is a major difference between the concept of paleontological species and those of biological and genetical species according to the acceptance in the Modern Synthesis. The concept of paleontological species considers the historical development of a species, a view that does not occur either in the genetic or biological species concepts. Despite this advantage given by the species historical development in time, many paleontologists continued to use morphospecies and typological species concepts on large scale. This situation can be explained by the large amount of data necessary in recognizing a paleontological species; in contrast, the morphospecies and typological species are more "practical": easy to define and use.

As our knowledge on the fossil record advanced it became possible to recognize various patterns in the history of life. One of them, namely the iterative evolution is of particular importance; according to it various morphologic features iteratively developed in the evolutionary history of a certain group of organisms. A new concept of paleontological species, named the *composite paleontological species* was defined by Georgescu and Huber (2009) in order to provide a method to recognize both the paleontological species and iterative evolution pattern: "A composite paleontological species is the basic unit with taxonomic significance in the fossil record, and has the following characteristics: (1) it is monophyletic; (2) it has a distinct range of morphologic variability, showing relative stability over a definable period of time and presenting relatively discrete evolutionary changes; (3) it is a morphologically heterogeneous and discontinuous entity, consisting of one or (mostly) more morphologic and/or paleoecological varieties; (4) it has its own and continuous developmental history traceable in space and time, which can be directly derived from the fossil record; and (5) its existence and integrity can be tested not only by comparative morphologic distinctiveness, but also by its response to paleoenvironmental and geological factors (e.g., paleoclimatic changes, sea-level fluctuations), as inferred from paleontology and related geological disciplines."

4.5 Speciation, Species Evolution, and Extinction

Species are natural entities, which exist for a definite period of time. There are three major processes in the history of a species: speciation, evolution, and extinction. *Speciation* is the process through which one species begins its evolution, and *extinction* is the process through which the species ceases its existence; the stratigraphic interval between speciation and extinction is also known as stratigraphic range. A species is unique in the fossil record; once it became extinct it will not reappear.

Species are morphologically heterogeneous complex units, with an uneven distribution of the individuals in space and time. Individuals of a species are clustered and form populations, groups of populations, races, subspecies, etc.; all these are *infraspecific categories*. The process of speciation happens when a population or group of populations becomes geographically isolated from the rest of the species and therefore, its gene pool can evolve independently. The populations that are geographically isolated are known as *allopatric*; two or more populations are referred to as *sympatric* when their areals of distribution present a distinct overlapping over a certain geographic area (Figure 4.15).

Species evolution can be studied in successions of samples; all the fossil species present morphologic fluctuations in time. Darwin (1859) considered that the species change throughout their existence, or stratigraphic range; this kind of species evolution is known as *gradualism*, or *phyletic gradualism*. Darwin (op.cit.) also mentioned a different perspective on speciation: "It is a more important consideration, leading to the same result, as lately insisted on by Dr Falconer, namely, that the period during which each species underwent modifications, though long as measured by years, was probably short in comparison with that during which it remained without undergoing any change." The process of speciation results in the life form diversity increase, and this is because the daughter species coexist with the parental species. There is a small number of species in which the evolution of the daughter species corresponds to the extinction of the parent species; such a case is referred to as *phyletic transition*.

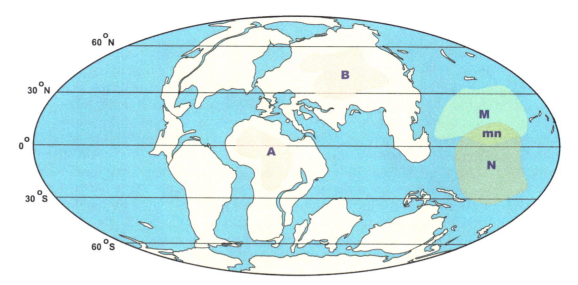

Figure 4.15 Idealized representation of allopatric populations (**A** and **B**) and sympatric populations (**M** and **N**, in which **mn** is the overlapping zone between them).

Recognizing the ancestor-descendant relationships between species is paramount in grouping them. Three kinds of groupings can be recognized according to how the species evolution matches the features considered taxonomically significant: *monophyly*, *polyphyly*, and *paraphyly* (Figure 4.16).

- *Monophyly* is grouping the species according to the direct ancestor–descendant relationships; paleontologists agree upon the fact that definition of monophyletic groups, which are considered natural, is preferable.
- *Polyphyly* often results in the formation of groups inconsistent with the ancestor–descendant relationships; the species are grouped according to one or more common features, which are often developed in different lineages. Notably, in the case of certain organisms such as some cnidarians, polyphyly is a frequent evolution process.
- *Paraphyly* is a kind of grouping that includes only a part of the descendant species from a common ancestor; paraphyletic groups are not consistent with the evolution process are at the best reflect it only partially.

4.6 Principles and Methods of Species Classification

There is more than one method to group species into higher rank units; for this reason all the specialists—biologists and paleontologists—studying the life form classification and the principles on which it is based agree upon the fact that the taxonomy of a group of organisms is not unique. Each organism presents a variety of morphologic features, but the use of one, a combination or their totality resulted in the definition of several currents of thinking among the taxonomists in the last 300 years: *essentialism*, *nominalism*, *empiricism*, and *evolutionary classification*.

- *Essentialism* postulates that all of the organisms in a group have the same essential features, which need revealing through scientific study, and it is the group essence

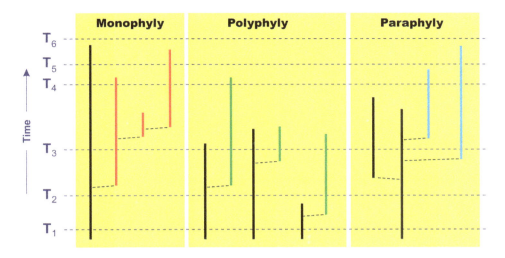

Figure 4.16 Schematic representation of species groupings; species grouped in the same higher category are given in red (monophyly), green (polyphyly), and blue (paraphyly).

which represents the basis for classification. Essentialist thinking led to the development of the *typological classification* or *typology*. A major impediment in typology practical application is the poor emphasis on the group variability, as the accent is put on the concept of *type*. Typology is also referred to as Linnaean classification, a method widely accepted in classification today throughout the biological and paleontological sciences; this method was for the first time developed in the Antiquity times by Aristotle of Stagira.

- *Nominalism* is a current of thinking in classification that considers that only specimens exist in nature and all the groupings of organisms at any level of classification are the product of human mind. This method was never used extensively but it was theorized upon in the eighteenth century.
- *Empiricism* is another current of thinking in organism classification that initiated prior to the publication of the *Origins of Species* in 1859. According to the empiricist current all the morphologic features are important in the organism classification, and therefore a successful classification must take them all in consideration. The method in itself was probably extremely accurate in identifying the organism groups, but a major impediment in the development of this current is that it could not provide a reason for the existence of the groups of organisms in nature. Empiricism was often criticized as offering a kind of objectivity nobody desires.
- *Evolutionary classification* was initiated by Charles Robert Darwin in the *Origins of Species* in 1859. According to the Darwinian Theory of Evolution species evolve from each other and it is possible to recognize the ancestor–descendant relationships between them by using a variety of data of paleontological and biological nature. Those taxa with significance in evolutionary classification should be defined by a combination of features that result from the common ancestry (resemblances) and as the result of evolutionary process that is divergent (differences).

Typological approach is overwhelmingly used in the paleontological classification today; its basic principle is the taxa grouping into units of higher rank by means of morphologic resemblance. This method is simple to apply and the taxa groupings are readily used in the current practice in paleontology and related sciences. However, the subjective factor is given by the differences in opinions between specialists on how close the morphologic resemblances should be in order to having a taxon included into one group or another. Another problem raised by a classification method based entirely on taxa resemblances is raised by a well-known pattern in the living world evolution, namely *iterative evolution*. The situation raised by the iterative evolution in typological classification is illustrated in the case of a small group of Cretaceous planktic foraminifera, which evolved radially elongate chambers with one bulbous projection at the distal end or a pointed termination. The test ultrastructure, which is a slowly evolving feature, indicates that the elongate chambers evolved independently in several lineages during the Cretaceous. Grouping the species with radially elongate chambers into one genus creates a polyphyletic grouping of species (Figure 4.17). Although the

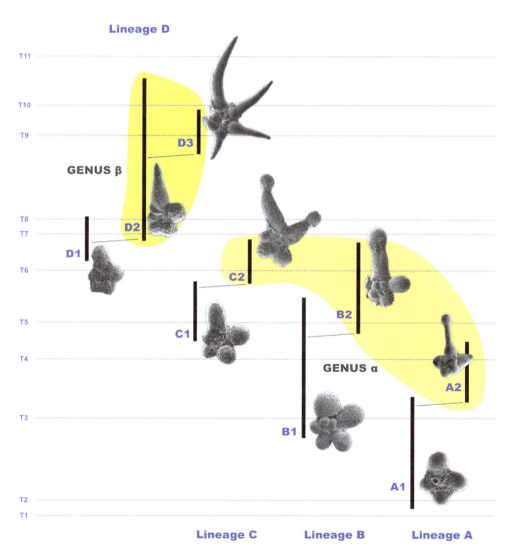

Figure 4.17 Species grouping into genera illustrated in the case of Linnaean classification. The two genera are recognized according to the chamber termination (tapering or with a bulbous projection).

genus can be described as an objective unit as all the species included in it bear the same feature, it cannot be used in evolution studies due to its polyphyletic nature. This is because the effects of iterative evolution are not included in the classification framework.

In evolutionary classification the principles are fundamentally changed when compared with typology. Species are grouped into lineages, which are the taxonomic units with significance immediately above the species level in evolutionary classification (Georgescu, 2009a, 2010). Therefore it can be demonstrated that each species with radially elongate chambers evolved from one that presents globular chambers throughout the ontogeny. Species grouping will take in consideration the test ultrastructure features and as a result, four distinct lineages are recognized (Figure 4.18). In this case, each lineage is described in a dynamic sense: species are grouped according to (i) common ancestry as shown by the almost identical test ultrastructure and (ii) achievements in the evolution

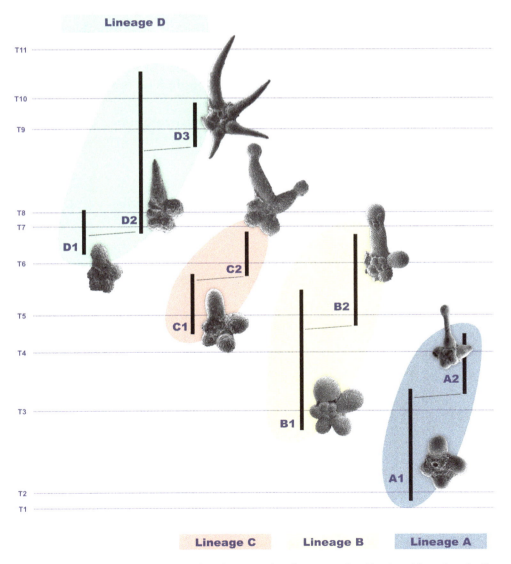

Figure 4.18 Species grouping illustrated in the case of evolutionary classification. Note that the lineage is the fundamental level of organization in this classification method.

process as demonstrated by the different chamber shape. In this case lineage definition is based on a combination of features that demonstrate both resemblances and differences between the species included; this sharply differs from the typological classification in which only the resemblances are taken into consideration.

Another major difference between the typological and evolutionary classification is the nature of the units above the species level. Whereas in typological classification the units of the same level are identical throughout the living world (e.g., a genus in bacteria has identical taxonomic status with a genus of reptiles), in evolutionary classification there are different kinds of lineages. Georgescu (2014) recognized four such kinds: *directional*, *branched*, *iterative*, and *condensed*, according to the general relationships between the species, mentioning also that new kinds of lineages can be defined as recognized through scientific observations (Figure 4.19). By this the evolutionary classification, in which the lineage is the fundamental unit of living matter organization, proves an opened system that can be improved as our scientific knowledge advances.

- *Directional lineages* are characterized by a continuous evolution between two or more features resulting in a monophyletic-linear succession of species.
- *Branched lineages* are characterized by the occurrence of at least one feature that presents divergent evolution, resulting in a monophyletic-branched succession of species.
- *Iterative lineages* are characterized by the repetitive evolution of one or a group of features, resulting in evolution of descendant species that considerably resemble each other and evolved from one ancestor.

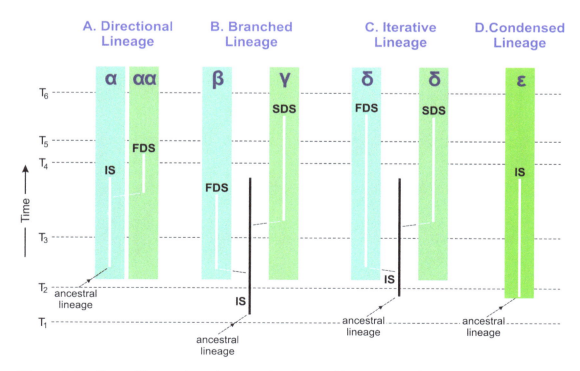

Figure 4.19 Types of lineages in evolutionary classification (Georgescu, 2014).

- *Condensed lineages* are characterized by a rapid evolution rate, sharp morphological differences when compared with the ancestral species, and relatively low morphological variability within the lineage.

A comparison between genera and lineages, the significant units immediately above the species level, shows an even more profound difference between the two classification frameworks. The typological taxonomic units are artificial in nature, identical throughout the living world; although they may be improved through the addition of new levels between the known ones, the units at the same level are identical. For this reason Georgescu (2011) noted the typological classification as having an axiomatic character. In contrast, the evolutionary classification has a scientific character, which is given in this case by the fact that new kinds of lineages can be defined.

4.7 Macroevolution

Macroevolution is a term used to describe the evolution of the taxonomic units of higher rank, usually above the species level. The evolutionary relationships between the higher categories are difficult to infer in general owing to the significant morphologic gaps that separate them. The difficulties are even higher in the case of the groups of organisms that do not develop skeletal hard parts, and therefore fossilize rarely. In such cases entire groups of organisms cannot be found in the fossil record; one of the most apparent result of the insufficient data generated by the fossilization bias is that the earliest group of a lineage that develops hard body parts would seem having a sudden evolutionary occurrence.

The ideal case in studying macroevolution is when the evolutionary transitions between higher categories can be documented at species level. For example, the planktic foraminifera were considered a monophyletic group, which evolved in the late Early Jurassic from benthic ancestors. It was also known that the planktic foraminiferal test presents two architectural types: serial and coiled. Species taxonomy and grouping at lower supraspecific levels (e.g., genus, subfamily, and family) were based on the fact that the serial type evolved from the coiled one much later, namely in the proximity of the Lower/Upper Cretaceous boundary. High-resolution studies showed by contrast to what was thought before, a morphologic transition between the benthic foraminifera as ancestors and serial planktics as descendants (Figure 4.20). The transition began with the benthic genus *Pleurostomella*, which evolved into *Praeplanctonia*, a genus with double chamber arrangement, triserial in the juvenile and biserial in the adult stage. *Praeplanctonia* evolved divergently into two descendant lineages formalized at genus level: *Archaeoguembelitria* (triserial chamber arrangement throughout) and *Protoheterohelix* (biserial chamber arrangement throughout). *Archaeoguembelitria* and *Protoheterohelix* independently developed the planktic habitat. Therefore, the planktic foraminifera do not form an evolutionary homogeneous group, but are a polyphyletic one, which consist of a mixture of several groups that developed the planktic habitat independently. This example shows that in order to provide reliable results macroevolution studies must be supported by evolution studies at species level (microevolution).

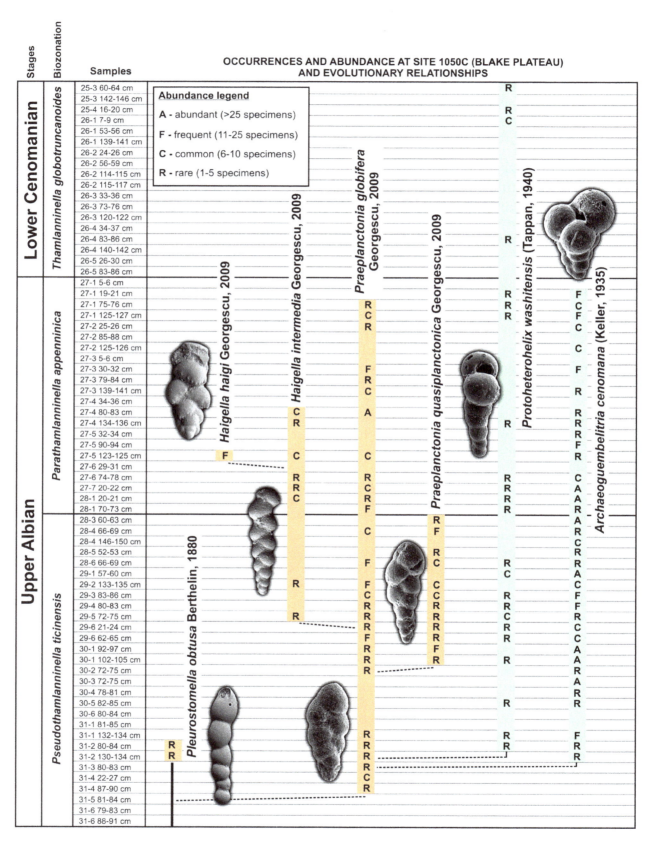

Figure 4.20 Example of evolution at species level that provides support for macroevolutionary interpretation (brown-benthic taxa, green-planktics); from Georgescu (2009b, fig. 3).

CHAPTER CONCLUSIONS

- A paleontological assemblage is a biased succession of biological populations.
- Organism variability is due to ontogenetic, genetic, and nongenetic causes.
- Ontogenetic variability is the result of organism growth during its lifetime; it can be isometric (shape remains unchanged) and anisometric (shape changes through ontogeny).
- Growth strategies include: accretion, addition, molting, modification, and mixed growth.
- Genetic variability is the reflection of the species gene pool; it can be either continuous or discontinuous.
- Nongenetic variability is induced by environmental factors (e.g., suprapopulation).
- Holotype is the name-bearer of a species; the holotype is selected only by the author of the species; paratypes are the remaining specimens from the original assemblage from which the holotype was selected.
- Neotype is selected when the holotype is lost, and preferably from amongst the paratypes.
- Lectotypes and paralectotypes are specimens selected by the first reviser from among the specimens originally illustrated by the species author in case the latter did not select a holotype and paratypes.
- Topotypes are specimens collected from the type locality of a species after the species formal description.
- Hypotypes are specimens collected from a different location and from the same or different stratigraphic level than the type level.
- Linnaean species (typological species) is the most frequently used species concept in paleontology; the concept of morphospecies is derived directly from that of Linnaean species.
- The first concept of paleontological species was given by G.G. Simpson (1961), and emphasized the role of the ancestor-descendant succession of populations and that of the factor time.
- The concept of composite paleontological species is derived directly and reflects the characteristics of the fossil record and data derived from it.
- Speciation is the process through which a species initiates its evolution; extinction is the process or event through which a species ceases its existence.
- There are four classification methodologies: essentialism, nominalism, empiricism, and evolutionary classification.
- Species are the fundamental unit in the typological classification (essentialism).
- In evolutionary classification the fundamental unit of organization of the life forms is that of lineage
- Macroevolution is the process of higher taxonomic category evolution; reliable results in macroevolution can be obtained only if they are supported by microevolution data.

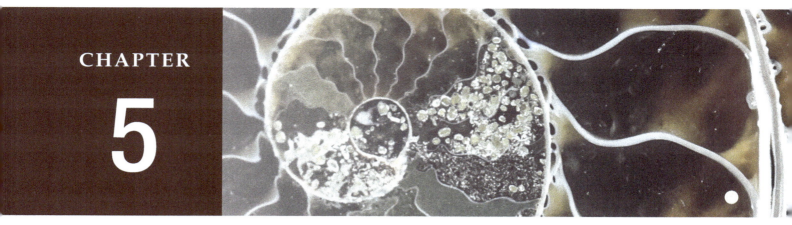

CHAPTER 5

LIFE EMERGENCE AND EARLY EVOLUTION

CONTENT

5.1 Living Matter Composition
5.2 Monomers and Polymers
5.3 Datasets in Deciphering Life Emergence and Its Early Evolution
5.4 Early Earth Events
5.5 Bacteria, Cyanobacteria, and Stromatolites
5.6 Banded Iron Formations
5.7 Evolution of Eukaryotic Cells
5.8 Emergence and Early Evolution of Multicellular Organisms
5.9 Evolution Evolves
5.10 Evolution of Burrowers and Earliest Shelly Fauna
5.11 Fossils of Burgess Shale

Chapter Conclusions

5.1 Living Matter Composition

The study of the elemental composition of the living organisms shows that there are a few dominant elements: carbon (C), hydrogen (H), oxygen (O), nitrogen (N), sulfur (S), and phosphorus (P) (Figure 5.1). The other elements occur in smaller amounts or as traces. This composition is relatively constant throughout the living realm. The remarkable homogeneity of the life forms on Earth in the elemental composition demonstrates that the successful life forms, which eventually led to the evolution of our species, evolved from a single primordial organism. These successful and evolving life forms on Earth are carbon-based. The recent discovery of life forms that have arsenic rather than phosphorus as major component in the genetic model demonstrates the possibility that the life forms evolved on our planet more than one time.

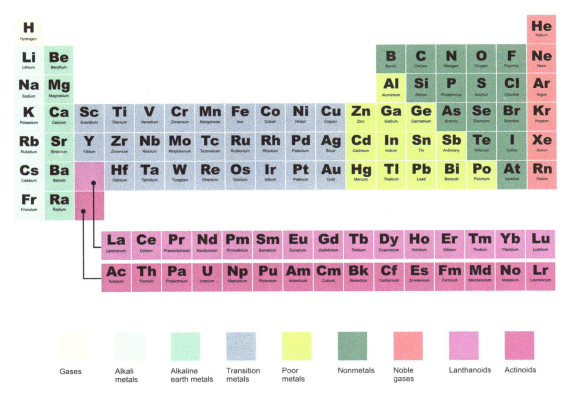

Figure 5.1 The periodic table of elements. Note that all the elements, which dominate in the life forms, are non-metals.

The persistence of the carbon-based composition of life forms on Earth can be demonstrated by studying the elemental composition of the substances resulted from the organic matter decay and transformation in the subsurface conditions. Natural hydrocarbons, such as the *natural gas, petroleum, asphalts, bitumen,* and *waxes* are formed through a succession of processes that begin after the death of organisms of various groups. Hydrocarbons began the process of formation through the accumulation of dead organic matter at the bottom of a sedimentary basin (e.g., lake, sea, ocean); the organic matter has two sources of provenance: from inside the basin (*autochthonous*) and transported from the adjacent continental regions (*allochthonous*); most of the organic matter accumulates in basins with *anoxic environments*, in which there is a layer of water enriched in dissolved toxic gases (CH_4, NH_3, CO_2, H_2S, etc.) in the proximity of the bottom of the basin. The dead organic matter that reaches the basin floor and starts the burial process represents about 5% of the organic matter total mass; about 95% of it is recycled through the trophic chains in the sedimentary basin. The organic matter decay begins before or immediately after its burial and various elements are expelled from its composition under the action of various bacterial groups and chemical reactions with the subterranean fluids and mineral environment. The process continues until only carbon and hydrogen remain from the original dead matter composition; hydrocarbons start to be expelled from the rocks that have significant amounts of dead organic matter in the following order: petroleum (between the temperatures of 60 and 120 °C) and natural gas, mostly methane (between the temperatures of 120 and 225 °C); the former temperature interval is known as the "*oil window,*" and the latter as "*gas window.*" Only carbon remains from the original dead matter at temperatures above 225 °C and hydrogen is the last expelled element; the carbon atoms are arranged in the crystal structure of the mineral *graphite*. The elemental composition of petroleum shows strong similarities to that of the modern living organisms, demonstrating the persistence of the organic matter composition through geological time (Figure 5.2).

Further significant information that can be used in the study of the life origin on Earth is provided by the molecular composition of the living matter. Most of the living organisms on Earth have their bodies dominated by water molecules (Figure 5.2); the cells of the human body have average water content of about 70% to 90%. Such

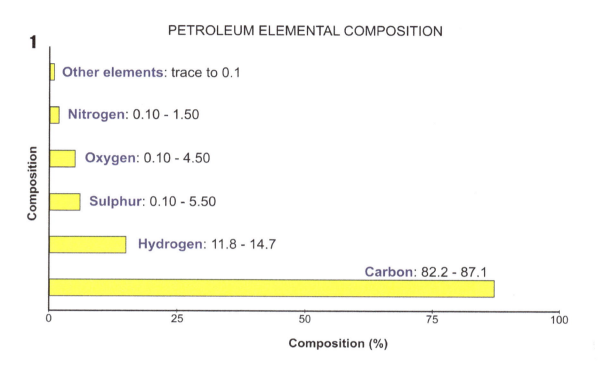

Figure 5.2 Petroleum elemental composition (above) and water molecules dominance in selected life forms after Schopf (1991, with modifications).

observation data are interpreted as demonstrating that life on Earth evolved in an aqueous environment; the two hypotheses on life origins today take in consideration that life emerged on Earth either in shallow water or in the deeper portions of the oceans.

5.2 Monomers and Polymers

It is evident in the rock and fossil record on Earth that the oldest rocks completely lack fossils and the oldest fossils are known from the upper part of Eoarchean; the oldest fossils are cyanobacteria, organisms that although prokaryotic, present an extremely complex structure at molecular level: thousands of molecule types, including about 500 RNA molecules. It is reasonable to assume that organisms of such complexity could not have been formed through simple reactions between atoms and simple molecules, and they rather evolved through a long process.

The dominant elements in life forms today, namely C, H, O, N, S, and P existed in the early Earth atmosphere in significant amounts; these elements are known as *biogenic elements* and their existence in the primordial cloud from which the Solar System was formed can be inferred from the chemical composition of the similar celestial agglomerations of matter in the Universe. The CHON-(SP) elements can easily combine to form simple, inorganic *monomers*, such as carbon dioxide (CO_2), methane (CH_4), ammonia (NH_3), water (H_2O), sulfur oxide (SO_3), etc. Such monomers have higher molecular mass when compared to those that dominated the terrestrial planet's primordial atmospheres: hydrogen (H_2) and helium (He); most of the light gases were removed from the terrestrial planet atmosphere by the solar wind, which initiated with the Sun ignition. As a result, the concentration in small monomers increased in the Earth's atmosphere, a phenomenon that further increased the possibility to react chemically and form larger molecular compounds.

A major problem in understanding the chemical reactions at the surface in the Earth early history was the chemical character of the atmosphere. An atmosphere such as the modern one, which is dominated by nitrogen (N_2) and had large amounts of molecular oxygen (O_2) could not have favoured the formation or larger organic molecules due to its oxidizing character. Alekxandr Ivanovich Oparin (1894–1989) provided around 1930 the first model in which, by contrast to everything that was believed before, the primordial atmosphere of our planet had a reducing character; further observations and experimental data confirmed Oparin's model. This model was primarily based on the observation that molecular oxygen could not have existed without its most important source, which is represented by the organisms capable of producing it through photosynthesis. Based on this argument Oparin further theorized that before the biological evolution there should have been on Earth periods in which the process of evolution happened at chemical and biochemical level. Organic molecules could have been accumulated at the Earth surface dissolved in water; the mixture of organic molecules was named *primordial soup* by Oparin.

The possibility that simple monomers can react chemically to form more complex molecules was demonstrated experimentally by Stanley L. Miller and Harold C. Urey in 1953. Miller and Urey considered that hydrogen was the dominant element in the early Earth atmosphere, and for this reason they started with a mixture of gases consisting of H_2, N_2, NH_3, CH_4, and H_2O; the source of energy consisted of electrical discharges and ultraviolet radiation. Seven amino acids were obtained in this experiment, with glycine (NH_2–CH_2–COOH) and alanine [CH_3–CH–(NH_2)–COOH] among them. Other organic substances that resulted in smaller amounts are formaldehyde (CH_2=O), hydrogen cyanide (HCN), acetylene (C_2H_2), cyanoacetylene (C_3HN), etc.

This experiment clearly demonstrated the possibility that small monomeric chemical compounds could have reacted at the surface of our planet forming more complex organic molecules. Additional experiments emphasized among others the role of CO_2 in the primordial atmosphere, a gas that was released in vast amounts by the volcanic activity in the early history of our planet. A major increase in the significance of this type of experiments happened when A. Macovei demonstrated in 1980 that higher temperatures are not necessary in the formation of simple organic molecules and therefore such chemical reactions could have happened at the environment temperature.

Simple monomers are frequent on the celestial bodies of the Solar System, most of the occurrences being known in the Jovian planets and their natural satellites. Simple organic molecules occur frequently outside the Solar System and they can be identified by studying the spectra of stars and agglomerations of matter: hydrogen cyanide (HCN), thio-formaldehyde (CH_2–S), methanol (CH_3–OH), ethanol (C_2H_5–OH), and among those organic substances recognized. Such occurrences indicate that formation of smaller organic molecules is a common process in the Universe and could have happened easily in the early history of the Earth.

The simple organic molecules can react chemically to form larger molecules, which consists of hundreds and thousands of atoms. Such large molecules are known as *polymers* and a polymer is formed through the repeating of a monomeric unit; the process through which organic monomers are chemically combined to form a polymeric molecule is known as *polymerization*. Polymers are frequent substances in the living world; an example of polymer is the cellulose that is formed by repeating the glucose monomeric unit ($C_6H_{12}O_6$). The process of polymerization commonly happens in two steps; it is exemplified in the case of simple amino acid molecules.

- Step 1. Dehydration condensation: two amino acid molecules combine to form a *dipeptide* molecule.
- Step 2. Dipeptide molecules combine to form a long polymeric chain known as *protein*; a protein is a *polypeptide*.

A similar polymerization mechanism happens in the case of other organic substances. For example two sugars combine through dehydration condensation to form a *disaccharide*; disaccharide molecules further combine to form a *polysaccharide* (carbohydrate). A more complex process is in the case of the nucleic acids where one sugar and one base combine to form a *nucleoside*; the nucleoside reacts with a phosphate to form a *nucleotide*; repeating nucleotide molecules result in the formation of a *polynucleotide* (*nucleic acid*) (Figure 5.3).

Additional chemical substances that came in the composition of the original soup had volcanic events as source. It is well-known from the study of the modern submarine vents that such phenomena enrich the surrounding waters with a variety of chemical substances, which could contribute to the formation of complex molecules. Moreover, the heat originating from the Earth interior created an environment around the vent in which the earliest life forms could have evolved. The submarine vents are considered today an alternative to the shallow water environments for the emergence of life on our planet.

The evidence of the chemical and biochemical phases of evolution are hard to find in the rock record mostly due to the fact that molecules do not fossilize. Despite this, an interesting discovery from the metamorphosed sedimentary rocks (3.8 to 3.7 billion years old) of the Isua Formation of southwestern Greenland comes to demonstrate the validity of our experiments and assumptions on the chemical and biochemical evolution of life on Earth. These rocks have a dark appearance, which suggest that they deposited in a reducing environment. Small-sized calcite minerals disseminated within

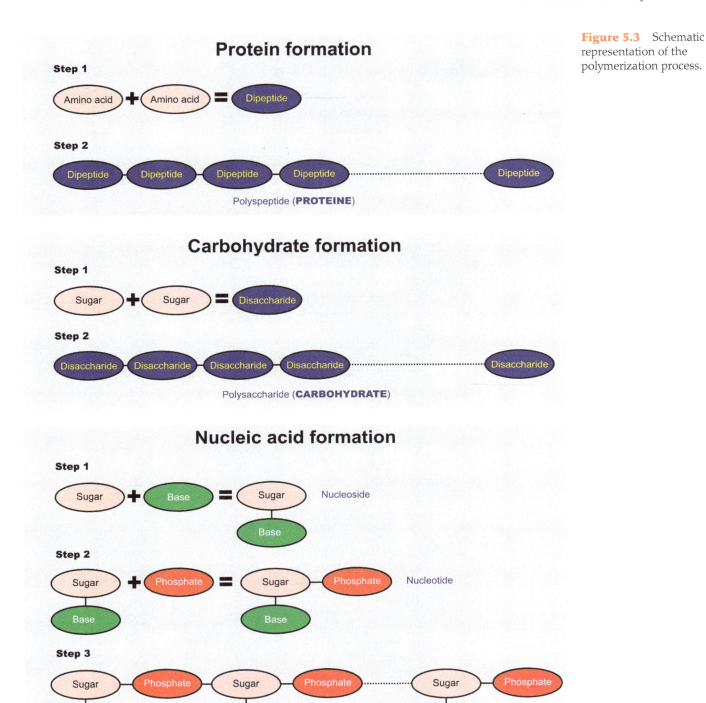

Figure 5.3 Schematic representation of the polymerization process.

the rocks demonstrate that molecular oxygen existed only in small amounts, so true layers of calcitic rocks could not form. Thin layers of graphite occur in these rocks; as the graphite is the final product of the organic matter transformation in the subsurface conditions during the burial process, the thin layers were interpreted as representing either the "fossilized" vestiges of the primordial soup or completely transformed early organisms, which were probably photosynthetic.

5.3 Datasets in Deciphering Life Emergence and Its Early Evolution

The difficulties in studying the emergence of life forms on Earth are raised mainly by the (i) small amounts of rocks with an age of 4 to 3.5 billion years old (when the life appearance happened) that survived the plate tectonic process, (ii) the high degree of metamorphic alteration of these rocks, and (iii) the fact that simple molecules do not fossilize. The separation between the continental crust and oceanic crust happened in the Earth early history and the modern plate tectonics beginning is dated in the proximity of the Archean/Proterozoic boundary; since then the oceanic crust, which floored the zones covered by oceans and in which the chemical reactions that led to the formation of the earliest life forms, was destroyed and reformed several times. Therefore, the simple observation followed by interpretation procedure we often use in the study of the fossil remains from younger sediments must be refined. There are five data sources that can be used in understanding the life origins on Earth: (i) rock record, (ii) fossil record, (iii) chemical record, (iv) experimental data, and (v) extrapolation of the biochemical and biological data known from the modern organisms.

- *Rock record* can provide data about the types of sedimentary rocks that occur in the stratigraphic record; the occurrences of these rocks can help in interpreting the chemical composition of the early Earth atmosphere and hydrosphere.
- *Fossil record* provides the direct evidence on the achievements in the life evolution. We must always take in consideration that the fossil record is not perfect and only a small number of organisms are fossilized; fossilization rate in the Archean and Proterozoic organisms is much smaller than that estimated for Phanerozoic organisms, due to the absence of hard body parts.
- *Chemical records* represent the occurrences of chemical fossils that can help in clarifying various aspects related to this topic. For example the $^{13}C/^{12}C$ ratio helps in recognizing the organic or inorganic origin of the rocks that yielded the analyzed isotopes.
- *Experimental data* are of paramount importance in recognizing the possible reactions between the inorganic and organic molecules during the chemical and biochemical evolution that preceded the life emergence. Experiments we realize today are helpful in producing models of the emergence of life and different of its components.
- *Extrapolation of biochemical and biological data* from the modern organisms can help in understanding the metabolic processes that powered the ancient life forms, morphologic components of the earlier life forms, as well as the probability of life occurrence on our planet.

5.4 Early Earth Events

The conditions of the emergence of life forms on Earth can be better understood if we study events in the early Solar System. The formation of the Solar System is explained by the *Solar Nebula Theory*. According to this theory, the Solar System formed from a cold cloud of dust and gases and this process started about five billion years ago. Initiation of the Solar System formation is a relatively late event in the history of the Milky Way, which was formed about one billion years after the Big Bang, the event that marks the initiation of the known Universe expansion (Figure 5.4). The beginning of the cold cloud rotation resulted in the dust and gases concentration in

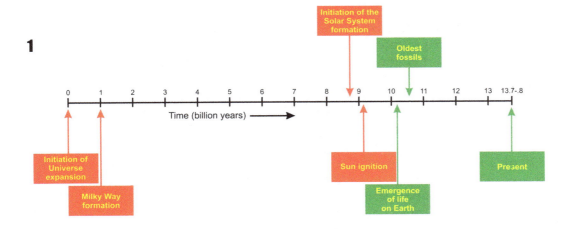

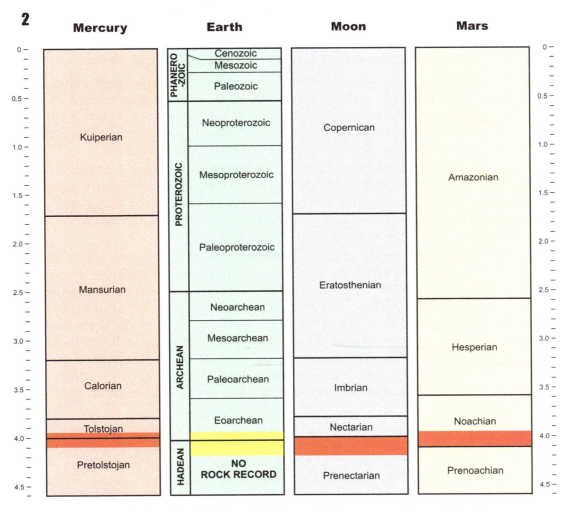

Figure 5.4 Time scale from the initiation of the known Universe expansion, marking the emergence of life on Earth and earliest fossils. 2: Timing of the catastrophic meteorite bombardment on Earth as inferred from the data from the Moon, Mercury, and Mars; stratigraphical zonation after Ogg et al. (2008).

the cloud's central portion and a number of concentrical rings around the center. Matter concentration led to the formation of the Protosun in the central part of the rotating cloud, and a number of *planetesimals* also known as *protoplanets* from which planets will form, gravitating around it; particle friction in the Protosun determined an increase of its temperature. Sun ignition began with the thermonuclear reactions in its mass; the beginning of the stellar phase in the Sun evolution was an event with dramatic impact for the early Solar System due to the energy it started to radiate and dissipate in the Solar System through the *solar wind*.

The formation of the Earth is dated circa 4.6 billion years ago. It is considered that the Earth's formation was influenced by gigantic impacts between the Proto-Earth and residual protoplanets; one of these collisions between the Proto-Earth and *Theia* resulted in the formation of the Moon. At the beginnings of its existence the Earth was a celestial body consisting of molten matter; its high temperature was due to the energy resulted from the collisions with the surrounding asteroids, planet embryos and planetesimals. Environmental conditions at the surface of the early Earth were quite different from those existing today.

Four major processes happened in the proximity of the early Earth surface and all of them influenced the environment in which life emerged on our planet.

- *Early crust* formation started due to the differences in the temperature between the hot early Earth and cold outer space; once the early crust was formed the heat remained trapped at the Earth interior. Temperature of the Earth's pore is approximately equal to that at the surface of the Sun. Volcanic phenomena were common at the early Earth surface during the period of early crust formation.

- *Early atmosphere* formed as a result of the mixture of gasses from two sources: those remained from the Solar System formation and new ones resulted through the outgassing process during volcanic phenomena. Notably, most of the lighter gasses such as hydrogen and helium were removed from the Earth early atmosphere by the solar wind initiated by the Sun ignition.

- *Early oceans* formation began early in the history of Earth through accumulation of water resulted from the vapour condensation. An additional source of water is represented by the comets, which have the nucleus consisting mostly of ice; these celestial bodies were more numerous in the Solar System about 4.5 billion years ago and their impact with the Earth resulted in the increase of the water amounts on our planet. Zircon minerals from Australia, which gave radiometric ages of about 4.4 billion years, and are the oldest known minerals on Earth required vast amounts of water to form; therefore, their occurrence is considered compelling evidence that oceans formed soon after the earth formation.

- *A catastrophic meteorite bombardment* is documented in the early history of the celestial bodies from the inner Solar System. The effects of this process can be seen at the surface of the Moon, Mercury, and Mars; the surface of these celestial bodies is not affected by plate tectonics and therefore, craters resulted from the meteorite impacts are still visible at their surface (Figure 5.5). This is not the case of the Earth where the plate tectonics obliterated the vestiges of this catastrophic series of events in the Solar System; however, we can infer from the data from the Moon, Mercury, and Mars that this meteorite bombardment happened in the interval between 4.2 and 3.95 billion years ago (Figure 5.4). The conditions at the Earth surface were extremely harsh during the meteorite bombardment and it is

Figure 5.5 Cratered surface of the Moon; many of the craters were produced during the catastrophic meteorite bombardment circa 4 billion years ago. © NASA/GSFC/Arizona State University; published with permission.

possible that entire oceans were vaporized during the impacts with bigger asteroids. Therefore, organic molecules formed several times and even life forms could evolve into more complex organisms only after the meteorite bombardment ceased.

5.5 Bacteria, Cyanobacteria, and Stromatolites

The oldest fossils on Earth were discovered in the *Apex Chert*; this lithological unit is part of a volcano-sedimentary unit of the Pilbara Craton of Western Australia; the rocks of this lithological unit consist of conglomerates that were transformed into chert through silicification. The age of Apex Chert conglomerates cannot be established by studying the fossil content due to the absence of a biostratigraphical framework for the Precambrian rocks; radiometric ages are completely unreliable in the case of these rocks due to the extensive metasomatosis process that fundamentally changed the conglomerate chemical and mineralogical composition. But the Apex Chert is stratigraphically situated between two lava flows, which yielded fresh minerals not affected by metamorphism that can be used in radiometric dating. The lava flow overlain by the Apex Chert is the Duffer Formation (about 3.47 billion years in age); Panorama Formation representing the lava flow that overlays the Apex Chert is about 3.46 billion years old. From these data it can be calculated that the Apex Chert age is of about 3.465 billion years (early Paleoarchean). The radiometric ages of the Duffer and Panorama Formations were calculated using the U-Pb method.

The fossils from the Apex Chert are small-sized filamentous bacteria, which were assigned to several genera among which *Primaevifilum, Archaeoscillatoriopsis, Eoleptonema,* and *Archaeotrichion*. The general cell shape and small sizes make these genera assignable to the bacteria and cyanobacteria, which are prokaryotic organisms; the smallest of them, namely *Eoleptonema* and *Archaeotrichion* are classified as bacteria, whereas the larger, which are assigned to *Primaevifilum* and *Archaeoscillatoriopsis*, are considered cyanobacteria. The occurrences of cyanobacteria in the Apex Chert are extremely interesting as the representatives of this division are photosynthetic organisms, which use the solar energy to bind energy as photosynthesized glucose. Although the Apex bacteria and cyanobacteria are recorded in a sedimentary rock that was totally transformed through metamorphism, their occurrences are indicative for the beginning of a major process, namely the increase of molecular oxygen amounts in the Earth atmosphere.

There is little useful information to bridging the gap between the Isua Formation "fossilized primordial soup" (3.8 to 3.7 billion years) and Apex Chert bacteria and cyanobacteria (3.465 billion years). It is highly improbable that the Apex fossils were the earliest organisms of their kind; most likely bacteria and cyanobacteria evolved earlier in the life history. This interpretation is apparently supported by the occurrence of small-sized calcite crystals in the Isua slightly metamorphosed sediments; there is a high probability that the molecular oxygen necessary for their formation was produced by already evolved photosynthetic organisms, such as cyanobacteria. In this case, the life forms on Earth could have evolved much earlier than we consider today and the successful colonization of our planet by isolated, microscopic organisms could have happened about 3.9 billion years ago, just after the catastrophic meteorite bombardment ceased.

The next step in the evolution of life on Earth is the occurrence of *stromatolites*, which can be described clusters consisting of billions of individuals of the bacterial and cyanobacterial groups. Stromatolites are large-sized (frequently with a dimension of 1 m) organo-sedimentary structures that grow through accretion and present at the surface a mucilaginous thin layer consisting of bacteria and cyanobacteria; most of the stromatolites are known as fossils, especially from the Precambrian rocks (Figure 5.6), and they are rare in the modern seas and oceans. The oldest stromatolites in the fossil record (circa 3.7 billion years old) are known from the Isua Formation (southwestern Greenland). Stromatolites present different shapes: domical, layered, conical, etc.; their shape is apparently controlled by the paleoenvironmental factors, such as water depth and energy (Figure 5.7).

The study of modern stromatolites showed that there are four components in its mass, and they are presented starting from the surface towards the centre.

- *Growth surface*: thin layer (~1 mm) consisting of oxygen-producing photosynthetic cyanobacteria and aerobic bacteria.
- *Undermat*: thin bacterial layer (~1 mm) consisting of non-oxygen-producing photosynthetic bacteria and bacteria that although are anaerobic, still can use oxygen -when available- during their life cycles.
- *Oxygen-depleted zone*: thicker layer (~1 to 2 cm) populated by anaerobic bacteria.
- *Stromatolite mass*: includes most of the stromatolite (from a few centimeters to >1 m) and consists of calcium carbonate; there are no living bacteria and cyanobacteria in this portion of the stromatolite, which can be simply characterized as being a biochemically produced rock.

The earliest stromatolites, namely those of the Eoarchean-Mesoarchean stratigraphical interval had a patchy distribution over the Earth surface, but starting in

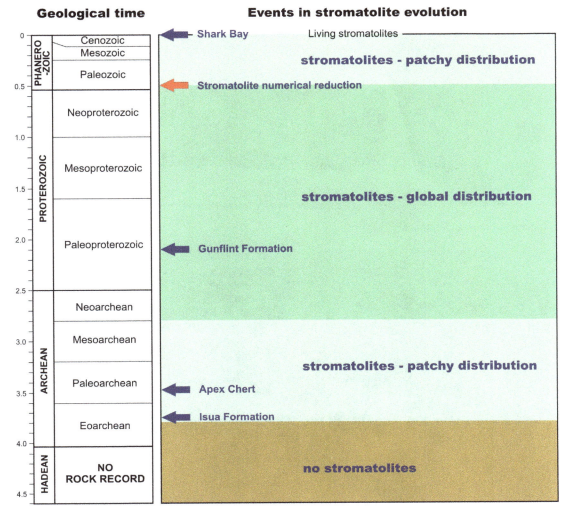

Figure 5.6 Major events in stromatolite evolution.

the Neoarchean they colonized almost entire surface of the Earth, occurring in the seas and oceans, coastal lakes, ponds, etc. Stromatolite global distribution continued in the Proterozoic and this setting persisted throughout the Proterozoic times; such a copious dominance of one group of organisms over the others, and for such a long period of time never happened in the Earth history and can be explained by the fact that for more than two billion years stromatolites did not have enemies. This setting changed dramatically just after the Proterozoic/Phanerozoic boundary when the stromatolites suffered a major numerical reduction generated by the evolution of the first shelly fauna and then by the molluscs, and especially the gastropods, both of which fed on stromatolites. Stromatolites are only rarely recorded in the Phanerozoic fossil record. The famous modern occurrence of stromatolites in the Shark Bay (Western Australia) is in a salty lagoon, a salinity constrained environment that is too harsh for the gastropods (Figure 5.7).

There is little variability in the stromatolite general characteristics from the Eoarchean to Quaternary; despite the differences in shape, stromatolites remain megascopic organo-sedimentary structures produced by bacteria and cyanobacteria. Stromatolite components are difficult to study due to their scarceness in the fossil record; most of the stromatolites do not preserve the bacterial and cyanobacterial

Figure 5.7 Example of fossil (1–2) and modern (3) stromatolites. 4: Living stromatolites in the Shark Bay (Eastern Indian Ocean, offshore Australia); photograph courtesy of Dr. C.M. Henderson (University of Calgary); published with permission.

components, which are easily destroyed during diagenesis and metamorphism in the absence of hard body parts. A distinct evolutionary pattern in stromatolites was observed with the study of the bacteria and cyanobacteria from the stromatolitic structures of the Gunflint Formation (southern Ontario, Canada), which are about 2.1 billion years old (Paleoproterozoic) (Figure 5.6). In contrast to the older representatives of the bacterial and cyanobacterial groups, namely those from Archean, the Gunflint prokaryotes present a remarkable diversity especially at genus level: *Eoastrion*, *Eosphaera*, *Animikiea*, *Kakabeckia*, *Huronipora*, *Gunflintia*, *Entosphaeroides*, etc.; this diversification also shows that new types of bacteria and cyanobacteria evolved and they are no longer restricted to the two primitive types: spherical and filamentous. Similar prokaryote assemblages re also recorded in other regions on Earth: North America (USA), Asia (Russia, India, and China), Australia, etc. These data document another major event in the history of life on Earth, namely the *prokaryote diversification*.

5.6 Banded Iron Formations

The evolution of photosynthesis is demonstrated by the fossil bacteria and cyanobacteria of the Apex Chert; this process resulted in the increase of the molecular oxygen in the Earth atmosphere. Photosynthesis evolved in organisms that developed their life cycles in marine environments, but the beginnings of this process are impossible to infer from the fossil record because older fossils than the stromatolites from Isua Formation are not known. The existence of oceans before the age of 3.5 billion years is demonstrated by the occurrences of ripple-marks and pillow-lavas. In addition, the zircon minerals of Australia, which are the oldest minerals on Earth (about 4.4 billion years old), require vast amounts of water to form.

Of paramount importance in recognizing the atmospheric conditions in the Earth early evolution is the rock record from the Isua Formation. Occurrences of small amounts of calcite ($CaCO_3$) indicate that CO_2 existed in the atmosphere approximately 3.8 to 3.7 billion years ago. But there is another kind of sedimentary rock in the Isua Formation that can cast additional light on the atmosphere composition; this rock type consists of an alternance of thin layers (millimeters to centimetres in thickness) of iron oxides, such as hematite (Fe_2O_3) and magnetite (Fe_3O_4) and layers of comparable thickness consisting of cherts and jaspers that present lower content of iron; the former have in general a reddish color and the latter are dominantly green to grey (Figure 5.8). These sedimentary rocks are termed *banded iron formations* (BIF).

Formation of the iron-rich layers in the banded iron formation happened during the volcanic events, when the iron expelled during volcanic eruptions is released in the atmosphere; the iron atoms are then combined with the dissolved oxygen from the oceanic water and such chemical reactions happened most likely in the uppermost layer of water in the oceans, where there is the highest concentration of dissolved oxygen. The source of the dissolved molecular oxygen is extremely important as it could not be released through outgassing during volcanic eruptions. The most probable source for the free molecular oxygen in the atmosphere is the photosynthetic process, as suggested by the modern settings. These indicate the existence of molecular oxygen in the atmosphere during the late Eoarchean times and indirectly the evolution of bacterial and cyanobacterial photosynthesis. Moreover, the banded iron formations through their precipitation mechanism appear a finely tuned marker for the atmospheric molecular oxygen levels. Whether the photosynthetic process occurred earlier on Earth is unknown and seemingly our knowledge is limited by the small amounts of surviving Eoarchean rocks.

The Archean and Proterozoic rock record yields additional information about the atmosphere and hydrosphere conditions in which sediments were formed at the Earth surface. *Pyritic conglomerates* are sedimentary rocks that contain significant amounts of the minerals pyrite (FeS_2) and uraninite (UO_2), two minerals which can be easily oxidized. Pyritic conglomerates occur in rocks of about 2.8 to 2.0 billion years age (Figure 5.8). The fact that such rocks and component minerals occur in the rock record without visible traces of oxidation indicates that the Earth atmosphere during the Archean and earliest Proterozoic had a reducing character as postulated in the model of Oparin. The molecular oxygen occurred in small amounts and represented most likely <1% of the total mass of the atmosphere.

Anaerobic photosynthesis was the dominant metabolic strategy among the Archean and Proterozoic bacteria and cyanobacteria and through it vast amounts of molecular oxygen were released in the Earth early atmosphere. In fact it is the only known mechanism through which the molecular oxygen could be produced at such

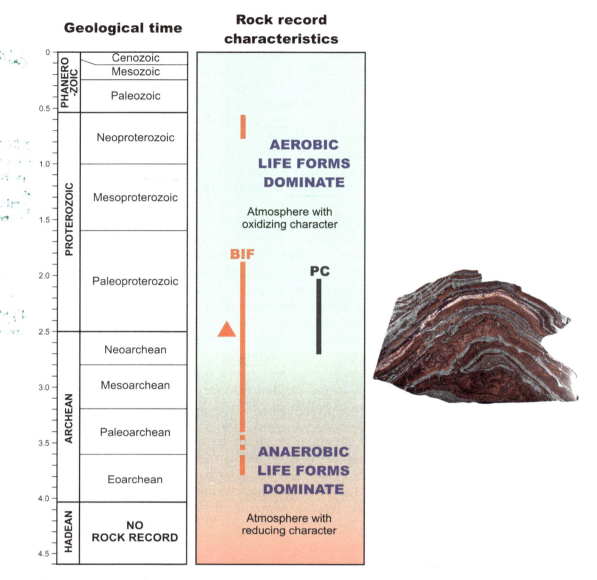

Figure 5.8 Banded iron formations (BIF) geological record and its relationships with the atmosphere character shift. Abbreviation: PC-pyritic conglomerate. Specimen of BIF from Illinois (USA), courtesy of Dr. R.O. Meyer (University of Calgary); published with permission.

higher rates. The numerical development of the stromatolites and significant diversity increase in the bacterial and cyanobacterial groups represent most likely the phenomena that triggered the considerable increase in the atmospheric molecular oxygen levels. The increase of molecular oxygen in the atmosphere is also demonstrated by the increasing occurrence frequencies of banded iron formations, which reached the maximum frequency approximately at the Archean/Proterozoic boundary, circa 2.5 billion years ago (Figure 5.8). Therefore, we can conclude that the life evolution produced a shift in the Earth atmosphere from *reducing* to *oxidizing*. Occurrences of banded iron formations are known also from the uppermost Proterozoic rocks of the Cryogenian-Ediacaran; formation of these younger banded iron formations is not related to the reducing-to-oxidizing atmosphere shift.

The oxidizing character of the atmosphere above the Archean/Proterozoic boundary is further demonstrated by other types of sedimentary rocks that started to accumulate on our planet, namely the limestones and so-called red beds. *Limestones* began to accumulate in vast amounts during the Proterozoic times after the atmosphere character shift from reducing to oxidizing; this pattern strongly contrasts with the patchy limestone distribution below the Archean/Proterozoic boundary. *Red beds* are sedimentary rocks that form mostly in continental environments in subaerial conditions and from this perspective contrast with the banded iron formations, which are subaqueous. Red beds present this dominant color due to the high content of iron oxides; in contrast to the banded iron formations the red beds contain in general less than 1% iron oxides in their composition. Examination of the red beds lithological components in thin sections reveals that the iron oxides do not occur as massive minerals, but mostly as coatings of the clasts of various nature (quartz, silts, and various lithoclasts); therefore, they resulted from the subaerial alteration of the sedimentary rocks such as sandstones and siltstones, another argument to demonstrate the existence of an oxidizing atmosphere.

The new oxidizing character of the Earth atmosphere, which was generated by the evolution and diversification of the prokaryotic organisms, changed completely the environment in which the life forms will evolve. It was the premise of the evolution of new and more complex types of life forms.

5.7 Evolution of Eukaryotic Cells

The formation of a stable oxidizing atmosphere about 2.0 to 1.8 billion years ago in the Paleoproterozoic created a new environment with new challenges for the living organisms. Single-celled diversifying prokaryotes were the only life to adapt to the newly defined environmental conditions. As a result, a new organism type evolved: the *eukaryotic* one. It is impossible at the present state of knowledge to recognize precisely the timing of evolution of the eukaryotes, especially because these life forms were single-celled and lacked mineralized organs that could readily fossilize. However, the eukaryotes were highly successful, producing a variety of morphologies and body plans during Proterozoic times. Ultimately they were the organism group that gave birth to the first multicellular life forms in the Late Proterozoic.

Eukaryotic cells can be briefly characterized by the fact they have a well-defined simple or multiple nucleus in the cytoplasm; this feature sharply contrasts to their prokaryote ancestor in which the nucleic substances are disseminated in the cytoplasm mass. But the eukaryote cell is more complex than the prokaryotic one especially by the existence within the cytoplasm of a number of distinct structures, also referred to as *organelles*, specialized for certain cell function:

- *Mitochondria* include the respiratory enzymes, but also act as the cell's "energy factory" by producing the adenosine triphosphate (ATP).
- *Ribosomes* are organelles with role in protein synthesis.
- *Plastids* (*chloroplasts*) occur only in the plant cells, and are those organelles used in the process of photosynthesis.

The high-resolution study of the organelles showed that their chemical compositions present similarities with certain eukaryotic groups. These resemblances apparently represent the key in understanding how the eukaryotes evolved. The fact that mitochondria and plastids are enclosed in their own membranes indicates that they originated either through symbiosis, or from engulfed prokaryotes that were not digested, but rather allowed to develop their life cycles inside the larger cell. The study of the cell components in various groups of living organisms resulted in a theoretical model on how different kinds of eukaryotic cell structures evolved. According to the model in use today, the process of eukaryote cell evolution happened in stages.

- *Stage 1.* Symbiosis between the anaerobic fermenting bacteria and flagellate bacteria; the resulting cell is the most primitive and also the simplest in the eukaryote realm.
- *Stage 2.* Symbiosis between the organism resulted from Stage 1 and bacteria capable to breathing oxygen; the later will develop in mitochondria. The newly formed eukaryote cell has the structure known in Kingdoms Fungi and Animalia.
- *Stage 3.* Symbiosis between the organism resulted from Stage 2 and bacteria capable of photosynthesis; the latter will develop into plastids (chloroplasts). This newly resulted eukaryote cell type is found in the representatives of the Kingdom Plantae.

The prokaryote and eukaryote cells are fundamental different not only from a morphologic point of view, but they also have different reproduction mechanisms. Cell subdivision in the prokaryotes is *asexual*, being often referred to as binary fission, budding, etc.; this is merely a replication process in which the anatomical and physiological characteristics are transferred from the parent cell to the offspring cells through simple multiplication. The rate of evolution in prokaryotes is very low, as the reproduction process can be mostly completed without mutations. Evolution of the eukaryotic cells resulted in the appearance of a new reproduction mechanism, which is the *sexual* one; the new life forms have their members separated in two distinct sexes, and the formation of an egg-cell from which a new organism can grow. The reproduction mechanism implies a combination of the gene pool from one male and one female individual; the combination process is not perfect and mutations frequently occur. Such mutations represent a major source of morphologic variability and experimentation in the course of the evolution process.

Finding the early Proterozoic eukaryotes in the fossil record is extremely difficult mainly because the molecules do not fossilize and therefore is almost impossible to find fossilized cellular nuclei. Despite this difficulty we were able to recognize true eukaryotes in the Proterozoic, and some of these are presented below.

- The oldest eukaryote evidence in the fossil record is dated at about 1.8 billion years, and is based on certain biomarkers (chemical fossils); no body fossil is associated to this occurrence.
- The first isolated true eukaryote cells were discovered in the cherts of the Bitter Springs Formation of the central Australia, and are dated at about 1.0 billion years. They are assigned to the red algae (rhodophytes) and green algae (chlorophytes, such as *Caryosphaeroides* and *Glenobotrydion*). Some of these were caught in the process of cellular division, and are among the very few cases in which even the nucleus was fossilized. In other (e.g., *Eotetrahedrion*) the offspring cell unique arrangement during the cellular division indicates clearly that the organisms belong to the eukaryote group; such arrangements are not found among the prokaryotes.
- *Bangiomorpha* is a red alga which first occurs in the fossil record in the Mesoproterozoic, circa 1.2 billion years ago; it presents multicellular filaments, which were

attached to the sea floor by a holdfast; the algal body was upright, and it represents probably the oldest known organism capable to orient its body towards the solar energy source. Species of this primitive alga exist even today.

- *Grypania* is a multicellular filamentous, ribbon-like alga recorded in sedimentary rocks as old as 1.4 billion years (Mesoproterozoic); its reported occurrence in sediments of Paleoproterozoic age still requires additional documentation.

- The most frequent eukaryotes in the Proterozoic belong to the acritarch group; their precise nature is unknown: red algae (rhodophytes), green algae (chlorophytes), dinoflagellates (single-celled plant-like protistan group), or other group of organisms. Large-sized acritarchs occur in the Neoproterozoic-early Paleozoic sedimentary rocks; the oldest occurrence is dated at about 1.0 billion years. A significant increase in their size happened about 850 million years ago, when for nearly 100 million years acritarch specimens with a diameter of about 1 cm occur in the fossil record (e.g., *Chuaria*). Acritarchs were planktic organisms and they document the eukaryote diversification and increased abundance in the Neoproterozoic times.

- A chlorophyte algal genus of the Neoproterozoic, which exists even today, namely *Torridonophycus* indicates that some eukaryotes were capable of developing protective cyst-like structures. Such structures helped the enclosed cell(s) to survive extremely cold or dry periods.

- *Melanocyrillium* first occurs in the Late Proterozoic as small-sized (about 60 nm) sack-like structures similar to those in the testate amoebas; the protective structure is mostly of organic nature, but it also contains agglutinated particles; the general shape of the test and its composition may indicate similarities with the primitive foraminifera.

The evolution of the Proterozoic eukaryotic cells was faster than that known in the prokaryotes; however, both groups share a relatively steady increase in diversity throughout the Proterozoic times. A faster pace of morphologic innovation and diversity increase will be achieved in the living world with the evolution of multicellular organisms.

5.8 Emergence and Early Evolution of Multicellular Organisms

There is compelling evidence about drastic changes in the Earth climate during the Late Proterozoic times. An ice-age began and continued for most of the Cryogenian, with the last pulse in the Ediacaran times; most of the surface of the Earth was covered under a thick sheet of ice. From the outer space the Earth approximately 600 million years ago might have looked similar to the Jovian satellite Europa; this period is informally referred to as the *snowball Earth*. The persistent low temperatures represented a formidable challenge for the life existence on our planet. Only the volcanic eruptions that penetrated through the ice sheet and released vast amounts of the climate-controlling gas carbon dioxide restored the ambient temperatures to values close to those existing before the glaciation. The oldest fossils belonging to multicellular organisms are recorded from the Cryogenian and have the earliest occurrence corresponding with the first glaciation pulse of the snowball Earth. The sponge *Otavia* is the oldest multicellular animal known on Earth up to date (Brain et al., 2012). Notably, algal multicellularity was achieved much earlier and is documented by the rodophyte *Bangiomorpha*.

The first multicellular animals from below the Proterozoic/Phanerozoic boundary were recognized in the Ediacara Hills from southern Australia; subsequent occurrences of fossils sharing the same morphologic characteristics were reported from around the world. The worldwide distribution of the Ediacaran organisms and high diversity of the body plans indicate that this group of organisms represented a successful adaptive radiation that colonized the shallow seas around the continental areas. Ediacaran fossils are preserved as impressions in a fine-grained meta-sandstone, the Rawnsley Quartzite. Ediacaran fossil communities are relatively frequent in the fossil record till the Ediacaran/Cambrian boundary.

Martin Glaessner (1906–1989) provided the first interpretation of the Ediacaran fossils. He noted that most of these fossils have affinities were medusa-like organisms, with circular shape and concentrical interior rings (e.g., *Cyclomedusa*, *Ediacaria*), other resemble the flat worms (e.g., *Dickinsonia*, *Ovatoscutum*, *Paleoplatoda*), and others the modern sea pens (e.g., *Charnia*). A smaller proportion of the Ediacaran fossils have typical segmented worm-like appearance (e.g., *Spriggina*). In addition, a number of Ediacaran genera are enigmatical fossils that cannot be included in any of the known phyla. Notably, none of the Ediacaran animals present exoskeleton or internal hard body parts (Figure 5.9).

Cephalization is well-documented in the case of *Spriggina*, where a cephalic shield evolved in the anterior part of the worm-like body. This pattern will be further developed in the history of life due to the animal need to protect the agglomeration of nervous cells, which will evolve into a brain and controls most of the body functions. A cephalic shield or a skull provides an increased protection against the predators and can increase significantly the chances of organism survival in the case of attack or injury.

The systematic position of the Ediacaran animals is a topic of scientific debate. Although Martin Glaessner originally considered that the Ediacaran organisms are the ancestors of the modern phyla, the absence of the internal cavities in all these organisms apparently indicates that they are more primitive than initially believed. The absence of internal cavities, and in addition sense organs (e.g., eyes, etc.), well-defined body parts (e.g., head, tail, limbs) are arguments to consider the

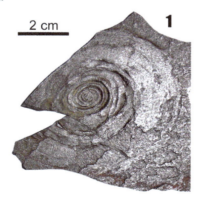

Tirasiana
Ediacaran,
Canada, Yukon

Tirasiana
Ediacaran,
Canada, Yukon

Figure 5.9 Examples of Ediacaran organisms. Both specimens from the paleontological collections of the University of Calgary.

Ediacara fauna consisting of organisms with primitive body plans. The feeding system of these organisms was fundamentally different when compared to the modern phyla, most which evolved in the Cambrian times. Absence of true digestive tubes indicates that nutrients were taken from the surrounding environments through *osmosis*; another possibility is that these organisms lived in *symbiosis* with photosynthetic algae that were similar with the zooxanthellae, a living mode similar to that of the modern cnidarians (corals). Seemingly such interpretations are supported by the absence in the Ediacaran assemblage of burrowing organisms, namely those that developed partially or completely their life cycles buried into the sediment.

The distinct characteristics, obvious primitiveness and low stratigraphic range were arguments to include the Ediacaran organisms into a distinct grouping, *Kingdom Vendozoa*. However, the consensus on their classification is not achieved yet. An overall evaluation of the Ediacaran fauna shows that it consists of organisms that despite their primitiveness when compared to the modern multicellular animals are extremely complex when compared to the single-celled eukaryotes, most likely their ancestors.

Three other metazoan occurrences in the Late Proterozoic demonstrate the accelerated evolution process, which was favoured by the development of the sexual reproduction mechanism: oldest trace fossils and exoskeletons. We do not know with precision the nature of the organisms that left the earliest trace fossils; these simple trails, which can be straight or curved are rather shallow, indicating that a worm-like organism produced them; however there are no traces of tracks left by appendages, so the organism must have had a very simple external morphology (Figure 5.10). The first organism with exoskeleton known so far is *Cloudina*, which was first discovered in the Upper Proterozoic sediments of Namibia (southwest Africa). *Cloudina* and genera with resembling morphology present a calcitic tube showing growth rings, and was probably secreted by a primitive cnidarian (Figure 5.11). The oldest endoskeleton also evolved in the late Proterozoic; the siliceous sponge spicules assigned to the cosmopolitan genus *Protospongia* also demonstrate the sponge evolution before the Proterozoic/Cambrian boundary.

'Worm' trail
Ediacaran,
Canada, Newfoundland

Figure 5.10 Example of Late Ediacaran worm trail. Specimen from the paleontological collections of the University of Calgary.

Figure 5.11 Examples of Ediacaran cnidarians from the Ediacaran of China. Photographs courtesy of Dr. Hong Hua (Northwest University, Xi'an, Shaanxi); published with permission.

5.9 Evolution Evolves

A comparison between the fossil record characteristics in the Archean-Proterozoic and Phanerozoic shows the existence of major differences between the organisms in the two stratigraphic intervals. Such comparison allows the identification of one pattern in the evolution mechanisms, which can be recognized only at the eon scale. The basic observation is that different organisms dominated the fossil assemblages in the Archean-Proterozoic and Phanerozoic: prokaryotes and eukaryotes, respectively. Prokaryotes and eukaryotes are fundamentally different in the cell organization, with the basic difference that the molecules and structures that transmit the hereditary features are disseminated in the cytoplasm mass in the former group, and concentrated in nucleus/nuclei in the latter. In addition, it is necessary to note that the Archean-Proterozoic encompasses nearly three quarters of the Earth history, whereas the Phanerozoic accounts only for circa one eighth of the Earth history.

5.9.1 Organism Nature

The oldest fossils on Earth are prokaryotes; they are stromatolitic structures of the Eoarchean Isua Formation and filamentous bacteria and cyanobacteria of the Paleoarchean Apex Chert. Stromatolites, which are organo-sedimentary structures built by bacteria and cyanobacteria gradually developed numerically, achieved global distribution in the Neoarchean and this pattern continued throughout the Proterozoic. Prokaryote diversification is the evolutionary process that apparently led to such a significant numerical development of the stromatolites. Eukaryotes evolved above the Archean/Proterozoic boundary, namely in the Paleoproterozoic when the rhodophytes evolved for the first time multicellularity in the algal group; the earliest multicellular animals occur in the Cryogenian times. The change in the dominant organisms happened in the early Cambrian, with the evolution of the first shelly fauna; apparently some of these gastropod-like organisms fed on stromatolites. The evolution of the first enemies represented the cause of the stromatolite numerical decline, and global distribution is reduced to a patchy distribution pattern.

5.9.2 Metabolic Strategy

Evolutionary transition from the prokaryotes to eukaryotes represented a major change in the metabolic strategies. The prokaryotes in the Archean-Proterozoic were anaerobic organisms; some of them could use the molecular oxygen when available, but the occurrence of atmospheric molecular oxygen was not a *sine qua non* condition in their life cycles. Eukaryotes evolved in the early Proterozoic after the content in molecular oxygen of the early Earth atmosphere increased significantly with the stromatolite numerical development. Practically the eukaryotes represent a response of the Earth early biosphere to the transition from the reducing to oxidizing atmosphere character. All the eukaryotes are aerobic organisms and this metabolic strategy will remain a constant throughout their evolution.

5.9.3 Reproductive Mechanism

Probably the most significant achievement in the evolution from prokaryotes to eukaryotes is the change in reproduction mechanism. Prokaryotes have a simple reproductive mechanism, which is based on cell multiplication; the reproduction mechanism at cellular level is based on the chromosome subdivision between the offspring cells and the DNA of offspring cells is identical to that of the parent cell. The reproduction mechanism is significantly changed in the eukaryote cells where the two sexes are separated; in this case formation of a new individual involves the combination of the male and female genetic information. The combination of the male and female chromosomes to form the egg-cell is often not perfect and errors occur, a process that results in a huge number of mutations, the raw material for evolution. The evolution of the sexual reproduction mechanism in the eukaryotes resulted in a major increase in the rate of morphologic changes among the organism groups in the tree of life. This organism feature is that that caused the significant diversity increase, which is a major feature of the Phanerozoic record in the history of life.

5.9.4 Ecologic Specialization

Prokaryotes are often generalist species, which can tolerate a wide range of variability of the environmental factors such as temperature, salinity, humidity, etc. Bacteria and cyanobacteria were widespread during the Archean-Proterozoic times and this was favored by their ecologic tolerance. Such ecologic tolerance pattern is further developed in the archaeans, also known as *extremophiles*; these organisms have the widest ecologic tolerance among the life forms on Earth. In contrast, eukaryotes are often specialized for certain ecologic niches and present much narrower margins of tolerance of the environmental factors. Directly derived from their tolerance to environmental factors is the population size in the two groups: prokaryotes develop usually large-sized populations consisting of a large number of individuals, whereas the number of individuals are considerably smaller in the eukaryote populations.

5.9.5 Evolution Tempo

This feature describes the rate at which new morphologic features and organism groups at species or above the species level evolve. Eukaryote evolution is well studied in the Phanerozoic sediments and Simpson (1944) recognized three tempos of the evolutionary process function of the species lifespan: *tachytely* (~1 million years), *horotely* (~10 million years), and *bradytely* (~100 million year). The prokaryote Archean-Proterozoic record shows a different pattern: species ranges are often longer than 100 million years documenting a much slower evolution tempo.

5.9.6 Evolution Characteristics

This refers to the major directions in which the evolutionary leaps are achieved. Prokaryotes present a wide variability with respect to the metabolic strategies; this contrasts to the small morphologic advances. Eukaryotes present constant metabolic strategies, which are either aerobic photoautotrophy and aerobic heterotrophy, and occur in algae-higher plants and animals, respectively. In contrast with the prokaryotes, eukaryote evolution happens frequently at tissue and organ levels; these organisms evolved thus body architectures with complex organs and systems specialized to perform certain body functions.

The overall changes in the proximity of the Proterozoic/Phanerozoic boundary with the change in dominance between the prokaryote and eukaryote organisms were for the first time reported by Schopf (2002) (Figure 5.12). The changes in the dominant organisms and evolution characteristics indicate that a profound shift in the life evolution on Earth happened about 540 million years ago; the change was from a slow evolutionary process with slowly evolving organisms in the Archean-Proterozoic to an extremely dynamic evolution process that results in the evolution of numerous specialized taxa. These data were used by Schopf (2002) to demonstrating that the process of evolution itself evolved in the life history on Earth.

Archean-Proterozoic (~ 3/4 of the Earth age)	Phanerozoic (~1/8 of the Earth age)
Dominant organisms: prokaryotes	**Dominant organisms:** eukaryotes
Metabolic strategy: anaerobic	**Metabolic strategy:** aerobic
Reproduction mechanism: asexual	**Reproduction mechanism:** sexual
Ecological specialization: generalist species	**Ecological specialization:** specialist species
Evolution tempo: very slow	**Evolution tempo:** extremely fast
Evolution characteristics: major innovations at metabolic level	**Evolution characteristics:** major innovations at tissue and organ levels

Figure 5.12 Changes in the characteristics of the life forms on Earth that document evolution's evolution (after Schopf 1999, simplified).

5.10 Evolution of Burrowers and Earliest Shelly Fauna

The beginnings of multicellular organism diversification during the Late Neoproterozoic did not involve only the evolution of endoskeletons (e.g., sponges) and exoskeletons (e.g., cnidarians). Relatively shallow traces left probably by worms or worm-like organisms also occur in the Late Neoproterozoic. Both of these two streams in multicellular organism diversification continued above the Precambrian/Cambrian boundary and diversified almost at exponential rate. Evolution of the endoskeleton and exoskeleton enforced the organism body and assured partial or total protection against predators. Burrowing into the sediment resulted in the first instance in a discovery of a new source of food, which is represented by the organic matter from the buried dead organisms.

Evolution of the burrowing organisms fundamentally changed the multicellular life forms on Earth. In the proximity of the Precambrian/Cambrian boundary the trace fossils became more numerous, thereby indicating that more and more organisms became adapted to such new ecologic niches. Burrower diversification resulted in a significant increase in the food amount that could be returned into the trophic chains. Dead organic matter in the superficial sediments became one of the richest food sources in the aquatic environments; this general setting is continued in the modern biocoenoses and was further developed in the terrestrial environments after the invasion of land in the late Silurian. The general appearance of the sedimentary rocks changed significantly in the proximity of the Precambrian/Cambrian boundary; this is the direct result of the evolution of burrowing organisms. Neoproterozoic rocks do not present traces of *bioturbation* because the Ediacaran animals soaked the nutrients at the sea bottom/water column interface or from the water column; bioturbation occurs occasionally in the latest Neoproterozoic and throughout the Phanerozoic. The process of evolution, diversification and consolidation of the burrowing faunas in the proximity of the Precambrian/Cambrian boundary is also referred to as the *agronomic revolution*.

Trace fossils in the latest Neoproterozoic and earliest Cambrian were relatively shallow thereby indicating that the organisms that produced them were mostly filter-feeders. Evolution of the predatory species determined a major change in the burrowing appearance, which became narrower and deeper (Figure 5.13). This indicated that some organisms started to use the superficial sediments to hide from their

Scolithos
Lower Cambrian,
Canada, Alberta

Rusophycus
Lower Cambrian,
USA, Ohio

Figure 5.13 Examples of Early Cambrian trace fossils. Both specimens from the paleontological collections of the University of Calgary.

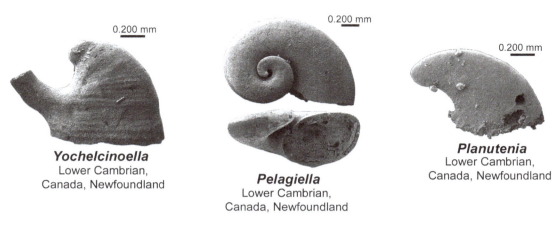

Figure 5.14 Examples of fossils of the Tommotian fauna. 1: Specimen illustrated by Skovsted and Peel (2007, fig. 4: D); © Acta Palaeontologica Polonica; published with permission. 2: Specimen illustrated by Skovsted and Peel (2007, fig. 4: I1-2); © Acta Palaeontologica Polonica; published with permission. 3: Specimen illustrated by Skovsted and Peel (2007, fig. 4: G); © Acta Palaeontologica Polonica; published with permission.

enemies. Moreover, species of long-ranging groups such as the inarticulate brachiopod *Lingula* that uses to live in dug galleries survived for a long time in the fossil record demonstrating the efficiency of the newly developed protection system.

The high rates of evolution of the trace fossils make them useful in the Cambrian biostratigraphy and are especially useful for the lower Cambrian sediments. The base of the Cambrian is defined with the aid of the first occurrence of the trace fossil species *Trichophycus pedum*, which was also known in the past under the name *Phycodes pedum*.

New organisms evolved in seas and oceans following the substantial increase in available food with the *agronomic revolution*. Diverse and new types of organisms are known from rocks as old as Early Cambrian; these fossils occur especially during the Tommotian stage and for this reason are often referred to as the *Tommotian fauna*. They are small (<5 mm in maximum dimension) and evolved shells with various chemical compositions (e.g., siliceous, calcitic, phosphatic, etc.) (Figure 5.14). The precise systematic position of most of the taxa of the Tommotian fauna is unknown, but the morphologic features of the mineralized skeleton indicate that some resemble molluscs, sponges, etc. Evolution and diversification of the Tommotian fauna largely correspond to the beginning of the stromatolite decline; most likely some of these taxa fed on stromatolites and represented the first enemies stromatolites had in more than 2.5 billion years of evolution.

5.11 Fossils of Burgess Shale

Burgess Shale is situated in British Columbia and is probably the most famous fossil lagerstätte in the world. Most of the fame and scientific significance of the Burgess Shale fossils is given by the extraordinarily good state of preservation of the soft-bodied organisms of middle Cambrian age. The fossils are found in fine-grained greenish-greyish-brownish shales. Sedimentation happened at a high rate, which explains the rapid burial process that led to the excellent preservation. Submarine sediment flows explain the absence of bioturbation in the rocks that yielded the fossils of Burgess Shale. Occurrence of short anoxic periods during the sedimentation further explains the conditions that resulted in the exquisite fossilization.

Extensive studies on the Burgess Shale fossils carried out over more than 100 years showed that the faunal diversity is extremely high and numerous taxa that cannot be included within any of the known phyla are frequent. Such unexpected diversity in number of taxa and plans of organization illustrates the process of diversification, which happened in the Cambrian period, following the "agronomic revolution" at the Proterozoic/Phanerozoic boundary. Some of the Burgess Shale fossils with high evolutionary significance are given below.

- *Anomalocaris* is the largest predator of the Cambrian seas; its success was partly due to the evolution of two big eyes that allowed spotting the prey easily. It was up to 1 m in length. The appendages without segmentation question its inclusion within the arthropod group (Figure 5.15).
- *Marrella* is the most frequent genus in the Burgess Shale fauna. It is an arthropod with vague trilobitomorph appearance; it has four strong backward-extending spines, two on the top of the body and two on the lateral sides.
- *Wiwaxia* is an enigmatical fossil; it is around 3 cm in length and the body is completely covered by small plates and protected by long and narrow spines. There is no apparent head.
- *Aysheaia* belongs probably to the phylum Onychophora, which includes worm-like organisms with multiple pairs of tube-like segmented legs. These

Figure 5.15 Examples of fossils from the Burgess Shale (British Columbia, Canada). All specimens from the paleontological collections of the University of Calgary.

morphologic features suggest that *Aysheaia* is probably an early descendant of a group with intermediate morphologic features between the segmented worms and arthropods.

- *Echmatocrinus* is a primitive echinoderm with the body protected by a number of small polygonal calcitic plates, which do not present a symmetrical arrangement. There are tentacles in the upper part of the body. The organism lived attached to the sea floor. Some authors suggested *Echmatocrinus* belongs to the octocoral cnidarians.
- *Pikaia* is one of the oldest cephalochordate known in the fossil record. It is around 5 cm in length and had an elongate body with distinct tail. The axial skeleton consists of notochord and the V-shaped muscles distribution is similar to that known in the living cephalochordate *Branchiostoma*.

CHAPTER CONCLUSIONS

- The dominant elements in the living organisms on Earth are: carbon (C), hydrogen (H), oxygen (O), nitrogen (N), sulfur (S), and phosphorus (P); they are also known as biogenic elements or CHON-(SP).
- The relative homogeneity in elemental composition indicates that all the life forms in the tree of life on Earth evolved from one primordial organism.
- The biogenic elements could combine in the early history of our planet to form simple inorganic monomers, such as carbon dioxide (CO_2), methane (CH_4), ammonia (NH_3), water (H_2O), etc.
- Alekxandr Ivanovich Oparin proposed the model in which the early Earth's atmosphere has a reducing character; organic molecules could not have accumulated in oxidizing environments similar to those existing today on Earth.
- In a famous experiment in 1953, Stanley L. Miller and Harold C. Urey demonstrated the possibility of organic molecule formation at the surface of the early Earth.
- Monomers can be combined to form polymers through the process of polymerization.
- There are two models that try to explain where the first life forms evolved on Earth: in shallower waters, and in deeper oceanic conditions in the proximity of the submarine vents.
- Isua Formation (about 3.8 to 3.7 billion years old) from southwestern Greenland contains the metamorphosed debris of the "primordial soup" and possibly earliest life forms on Earth; however, if present these early organisms are beyond recognition due to the high degree of metamorphism affecting the rocks.
- The most important events in the early Earth history were the early crust formation, early atmosphere formation, ocean formation, and a catastrophic meteorite bombardment that happened about 4.2 to 3.95 billion years ago.
- The oldest isolated fossils are known from the Apex Chert (about 3.45 billion years old) of Western Australia; they belong to the bacteria and cyanobacteria groups. They also demonstrate the evolution of photosynthesis.
- The first living stromatolites were discovered in a salty lagoon at Shark Bay in Western Australia.
- The oldest well-documented stromatolites are known from the Eoarchean Isua Formation (3.7 billion years old).

- Stromatolites achieved global distribution in the Neoarchean times.
- Banded iron formations (BIF) demonstrate the increase of molecular oxygen in the Earth atmosphere. The climax in these rock accumulations is dated about 2.5 billion years ago, in the proximity of the Archean/Proterozoic boundary.
- Limestones and red beds, which formed frequently in the Proterozoic and Phanerozoic, demonstrate the reducing-to-oxidizing shift in the earth atmosphere character about 2.0 to 1.8 billion years ago.
- Eukaryotes differ from the prokaryote ancestor mainly by having a well-defined nucleus in the cytoplasm. Eukaryotes developed a sexual reproduction mechanism.
- The oldest multicellular organisms evolved in the Paleoproterozoic and belong to the rhodophyte algal group. The oldest multicellular animals evolved in the Late Proterozoic (Cryogenian).
- Ediacaran animals are considered primitive due to the fact they lack internal cavities, sense organs (e.g., eyes, etc.) and well-defined body parts (e.g., tails, head, limbs, etc.).
- Organisms with exoskeleton (e.g., *Cloudina*) and internal skeleton (e.g., *Protospongia*) evolved in the Late Proterozoic.
- The comparison between the characteristics of the dominant organisms in the Archean-Proterozoic and Phanerozoic demonstrates that the process evolution evolved since the life emerged on Earth.
- Trace fossils became abundant and diverse in the proximity of the Proterozoic/Phanerozoic boundary.
- Beginnings of burrowing increased the food amount in seas and oceans, therefore contributing to the life diversification in the early Cambrian; this process is known informally as the "agronomic revolution."
- Burgess Shale of the middle Cambrian of British Columbia is probably the most famous fossil lagerstätte.
- The fossils of Burgess Shale document the process of life diversification during Cambrian times.

CHAPTER 6

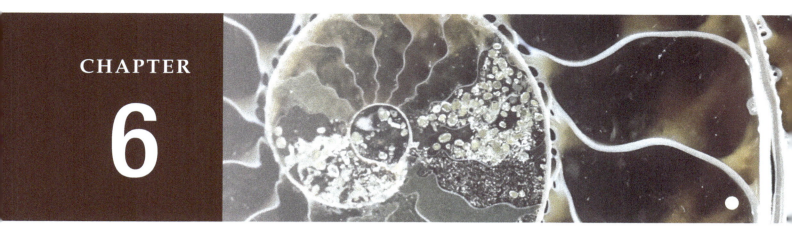

INVERTEBRATE EVOLUTION

CONTENT

6.1 Poriferans
6.2 Cnidarians
6.3 Bryozoans
6.4 Brachiopods
6.5 Monoplacophorans, Polyplacophorans, and Scaphopods
6.6 Gastropods
6.7 Bivalves
6.8 Nautiloid Cephalopods and Closely Related Major Groups
6.9 Ammonoid Cephalopods
6.10 Coleoid Cephalopods
6.11 Trilobites
6.12 Crustaceans and Chelicerates
6.13 Echinoderms
6.14 Graptolites

Chapter Conclusions

6.1 Poriferans

The representatives of this group were rarely studied from an evolutionary perspective despite their overall importance as the most primitive living group of multicellular organisms. This is partly due to the relatively rare occurrences of well-preserved complete organisms that also have the ultrastructural features in good state of preservation. Another difficulty in reconstructing poriferan evolution is represented by the fact that most of the occurrences are known from micropaleontological samples

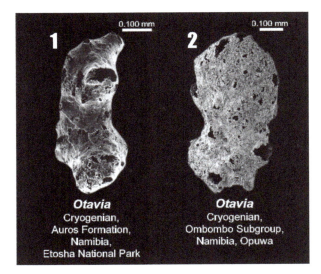
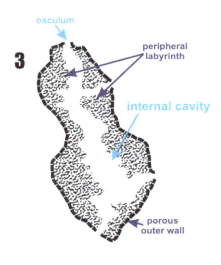

Figure 6.1 The oldest multicellular organisms: poriferan *Otavia* of the Cryogenian times. 1: Specimen illustrated by Brain et al. (2012, fig. 3: a); © South African Journal of Science. 2: Specimen illustrated by Brain et al. (2012, fig. 3: c); © South African Journal of Science. 3: Schematic representation of the body architecture of *Otavia* (Brain et al. (2012, fig. 6); © South African Journal of Science.

as isolated skeletal elements (microscleres and macroscleres) that cannot be used in reconstructing the whole skeleton; in fact, wherever such sclerites occur in micropaleontological samples it is impossible to assess if they were part of one or more poriferan animals.

The oldest known fossil poriferan is *Otavia*, which was described from the Cryogenian (Neoproterozoic) rocks of southern Africa (Brain et al., 2014) (Figure 6.1). It is a small-sized genus, often microscopical, with irregular outline and gently undulated surface with small-sized circular or elliptical openings. The larger of these openings are probably *osculi* through which water was expelled from the internal cavity; the smaller openings were probably inhalant pores through which the water current carrying nutrients and dissolved oxygen entered the central cavity. Stratigraphically *Otavia*'s earliest occurrence is in the proximity of the first pulse of the succession of glaciations that marked the Snowball Earth. This poriferan genus is not assigned to any of the known classes of the phylum Porifera, but probably belongs to class Calcarea, which includes sponges with calcareous ($CaCO_3$) spicules; from a mineralogical perspective the spicules can consist of either calcite or aragonite. Class Calcarea occurs in the fossil record throughout the Ordovician-Holocene stratigraphical interval, possibly earlier in the Cambrian times; probably one of the most spectacular set of occurrences is from the Upper Cretaceous chalks of Europe (Figure 6.2). A continuous stratigraphical record between *Otavia* of the Cryogenian-Ediacaran and calcareous sponges of the Cambrian-Holocene, Ordovician-Holocene does not exist; therefore, the two can be unrelated from an evolutionary perspective.

Class Hexactinellida includes taxa with siliceous skeleton consisting of spicules that intersect at angles of 90°. Mostly isolated spicules and rarely well-preserved skeletons with spicules in anatomical connection occur in the Neoproterozoic (Ediacaran); hexactinellid sponges occur throughout the Cambrian-Holocene stratigraphical interval (Figure 6.2). Most of the modern sponges are included within the class Demospongea that occurs in the Cambrian-Holocene stratigraphical interval, possibly in the Neoproterozoic (Ediacaran). Demosponges have the skeleton consisting of

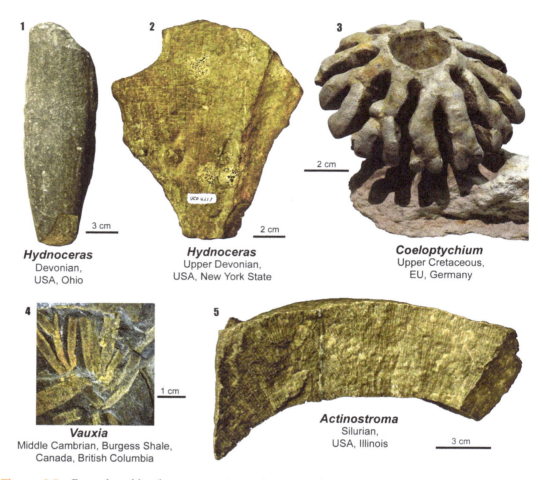

Figure 6.2 Examples of fossil sponges. 1–2, 4–5: Specimens from the paleontological collections of the University of Calgary. 3: Specimen from the Senckenberg Natural History Museum, Frankfurt; published with permission.

organically precipitated silica or an organic substance named spongin, or a combination of the two; in contrast with the hexactinellids, in demosponges the spicules intersect at angles of 60°/120°.

Class Stromatoporata includes some of the most important reef-builders of the Paleozoic times, namely from the Ordovician-Devonian stratigraphical interval. They are informally known as stromatoporoids and for a long period their systematic position was enigmatical. They present a laminated internal structure (Figure 6.2). The stratigraphical range of this class is Ordovician-Cretaceous; notably, sponges that have considerable resemblances with the stromatoporoids are also known from the modern faunas of reefal environments.

Class Archaeocyatha occurs only in the Cambrian, with the highest diversity and most occurrences in the Lower Cambrian; they are in decline during Middle Cambrian and only one occurrence is known from the Upper Cambrian. Most of the archaeocyathids are small, with the cup diameter of 1 to 3 cm, only rarely larger; the largest archaeocyathid has the cup diameter slightly over 50 cm. Archaeocyathids were benthic organisms and developed the life cycles permanently attached to the sea floor with the aid of a root-like structure. The cup-like body of an archaeocyathid specimen consists of two walls separated by an empty space (*intervallum*), which define an empty central cavity; the two walls (inner and outer) are connected by

Figure 6.3 Examples of archaeocyathid sponges. 1: Specimen from the Museum of Natural Sciences, Paris; photographed by the author, August-2014. 2: Specimen from the paleontological collections of the University of Calgary.

septa (Figure 6.3). All the calcitic structures of an archaeocyathids, excepting for the root-like structure are porous indicating that these organisms were filter-feeders. The representatives of this sponge class were the first invertebrates that built reefs in the history of life; such reefs were small when compared with the modern reefal structures: a few meters in length and with a height of maximum 0.5 m.

6.2 Cnidarians

Little is known about the origins of cnidarians. Evolution of *Cloudina* and other taxa with resembling morphologies (e.g., *Sinotubulites*, *Conotubus*, etc.) in the Ediacaran times represent an early evolution pulse of the cnidarian group. There is no continuity in the fossil record between these earlier cnidarians and the first major diversification of the group, which is documented in Ordovician. The study of fossil cnidarians is particularly important partly because the representatives of this group contributed massively to the construction of reefal structures for the most part of the Phanerozoic. Most of the reef-building cnidarians are included within the subclass Zoantharia of the class Anthozoa. The main orders of zoantharians from the perspective of the importance in the fossil record are: Tabulata, Rugosa, and Scleractinia.

Order Tabulata is known from the Ordovician-Permian stratigraphical interval. It is one of the two dominant Paleozoic zoantharian orders, and became extinct at the major crisis in the history of life at the Permian/Triassic boundary. All the tabulate corals are *colonial*; solitary taxa are not known among the representatives of this order. Each individual in the colony is known as *corallite* and has the interior subdivided by horizontal plates or *tabulae* (singular: tabula) (Figure 6.4). This type of corallite subdivision contrasts sharply with those known from the other cnidarian orders where the intra-corallite skeletal structures are only or mostly longitudinal. Corallite distribution within the colony varies from tightly packed without empty spaces between them (e.g., *Favosites*) to aligned forming chain-like structures (e.g., *Halysites*, *Catenipora*) or completely isolated (e.g., *Syringopora*).

Order Rugosa evolved in the Ordovician and became extinct at the Permian/Triassic boundary, and together with the representatives of the order Tabulata

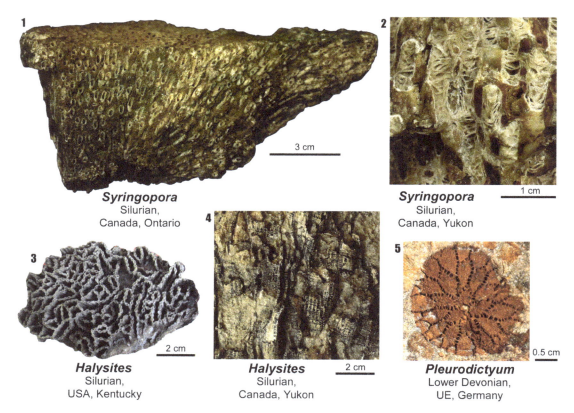

Figure 6.4 Examples of stony corals of the order Tabulata. All specimens from the paleontological collections of the University of Calgary.

were the most important reef-builders in the Paleozoic. Zoantharians of the order Rugosa are also known as tetracorals and include colonial and solitary taxa (Figure 6.5). The interior of the one corallite is subdivided by septa, which are longitudinal structures that can merge in the central portion of the corallite forming a *columella*. The septa distribution shows a four-fold symmetry, hence the name of tetracorals (Figure 6.6). *Cardinal septum* is an isolated structure and no other septa are attached to it; *counter septum* is situated in diametrically opposed to the cardinal one; *counter lateral septa* are symmetrically situated on each side of the counter septum and attached to it; two *alar septa* are symmetrically arranged between counter laterals and cardinal septum; all the other septa are considered minor and are added in sets during ontogeny. Horizontal structures further subdivide the corallite interior: tabulae and dissepiments. *Order Heterocorallia* is one small group of Late Paleozoic (Carboniferous) that probably evolved from order Rugosa (Figure 6.6).

Subclass Zoantharia was one of the most affected groups at the Permian/Triassic boundary. All orders of Paleozoic zoantharians became extinct during this mass extinction during this major crisis in the history of life. No cnidarians with mineralized skeleton occur in the Early Triassic (Induan-Olenekian), and the group began to develop again mineralized skeletons in the early middle Triassic (Anisian).

Mesozoic corals present a different type of septa distribution when compared to the Paleozoic corals and completely lack tabulae. Based on these significant morphologic differences they are included in a distinct order: *Scleractinia*; they are also referred to as hexacorals, a name which is derived from the six-fold symmetry

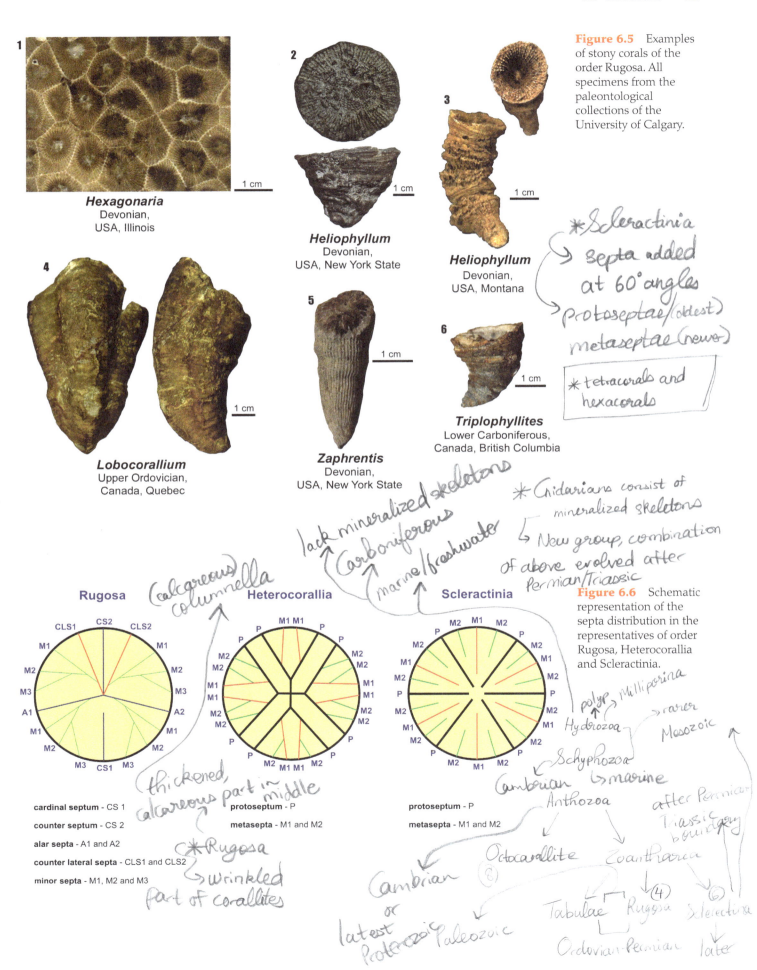

Figure 6.5 Examples of stony corals of the order Rugosa. All specimens from the paleontological collections of the University of Calgary.

Figure 6.6 Schematic representation of the septa distribution in the representatives of order Rugosa, Heterocorallia and Scleractinia.

recognized in septa addition and distribution (Figure 6.6). Scleractinians are the only zoantharians in the Mesozoic and Cenozoic. Their septa present a first cycle of *protosepta* or primary septa that are situated at an angle of 60° of each other. The following cycles of secondary septa (*metasepta*) are added between the previous ones and their number increase as follows: 6 in the first set, 12 in the second set, 24 in the third one, etc. Scleractinian corals are solitary or colonial (Figure 6.7) and were major reef-builders during the Mesozoic and Cenozoic, alone or mostly in association with other organism groups, such as rhodophytes, chlorophytes, sponges, and bivalves. A major feature in the scleractinians is that a significant part of the species in this group

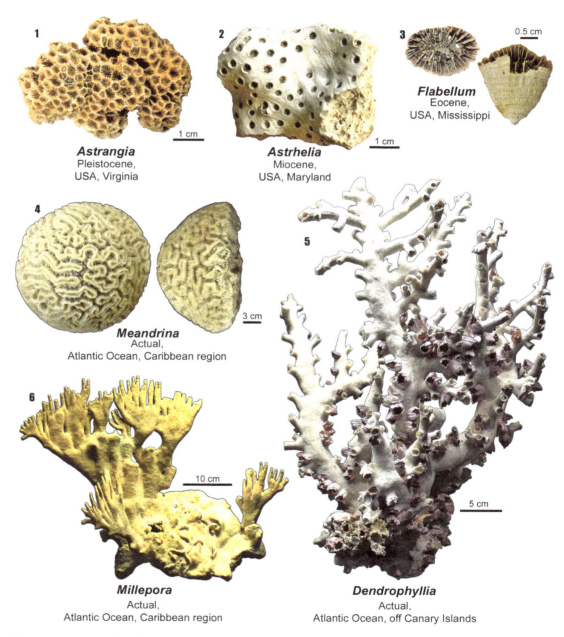

Figure 6.7 Examples of stony corals of the order Scleractinia (1–5) and cnidarians of class Hydrozoa (6). 1–4: Specimens from the paleontological collections of the University of Calgary. 5–6: Specimens from the Senckenberg Natural History Museum, Frankfurt; published with permission.

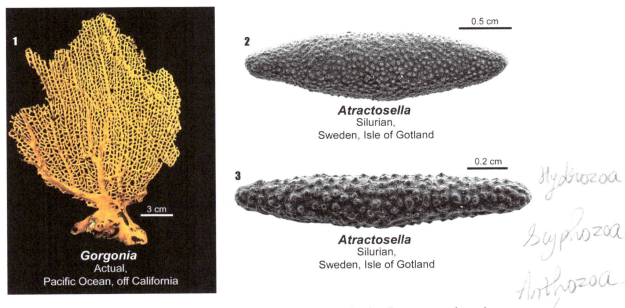

Figure 6.8 Examples of octocorals: colony (1) and fossil sclerites (2–3). All specimens from the paleontological collections of the University of Calgary.

live in symbiosis with photosynthetic algae (zooxanthellae); according to the occurrence of such algae, scleractinians can be *hermatypic* and *ahermatypic*.

- *Hermatypic scleractinians* live in symbiosis with the zooxanthellae and for this reason are restricted to the shallow, warm and clear waters, which allow the solar light to penetrate; this type of corals can attain large sizes and dominate in the reefal environments. They are distributed down to a water depth of about 90 m, but most of them live above 50 m.
- *Ahermatypic scleractinians* live without the symbiosis with zooxanthellae; therefore their bathymetric distribution is not restricted to the shallow and warm waters. The representatives of this informal group colonized the oceanic realm down to a depth of about 5000 to 6000 m, and can live in much colder waters than the hermatypic corals.

The other subclass of Anthozoa is *Octocorallia*, which includes among other subgroups the sea pens and sea fans. Individuals within colonies present in general an eight-fold symmetry. In the fossil record they occur mostly as microscopical calcareous sclerites in which the skeleton is disintegrated after organism's death (Figure 6.8). The sclerites occur rarely in the Ordovician-Holocene stratigraphical interval.

The other two classes of phylum Cnidaria, namely *Hydrozoa* and *Scyphozoa* are rarer in the fossil record. One special case is that of the hydrozoans of the *order Milleporina* (Figure 6.7); these colonial organisms present calcareous colonies and occur relatively frequently during the Cenozoic times, when they contributed significantly to reef-building.

6.3 Bryozoans

The representatives of this phylum of lophophorates evolved in the Ordovician and occur abundantly in the modern environments, especially the marine ones. No solitary taxa are included among bryozoans, which are only colonial. The colony, which is termed *zoarium*, consists of small-sized, microscopical individuals known as *zooids*; there can be tens to millions of zooids within one zoarium. Bryozoan colony is mostly attached to the substratum with a root-like structure but taxa that are unattached or intermittently attached are also known. The soft body parts of the different individuals are protected by the rigid structure of the calcareous colony and in many taxa they evolved a lid-like structure termed *operculum* that helps in completely enclosing and therefore, protecting the individual when retracted. The representatives of the phylum Bryozoa began their evolution as marine organisms but in the Mesozoic and especially Cenozoic colonized fresh and brackish water environments.

Most of the fossil bryozoans are included in two classes, namely Stenolemata and Gymnolemata, each of them represented by more than 600 genera; both classes occur in the same stratigraphical interval: Ordovician-Holocene, but their evolutionary history differs considerably. *Class Phylactolemata* occurs sparsely in the fossil record due to the lack of a calcified colony (Neogene-Holocene, with questionable occurrences in the Upper Cretaceous); less than 15 genera are included within this class, and all of them are freshwater taxa.

Class Stenolemata presents the classical bryozoan architecture; in general they are non-operculate. Colonies are well-calcified: most of the taxa are calcitic, only rarely aragonitic. Stenolemates are marine organisms, and dominated numerically phylum Bryozoa throughout the Ordovician-Cretaceous stratigraphical interval; a decline in diversity is recorded during Cenozoic times. The representatives of three stenolemate orders occur frequently in the fossil record. *Order Cyclostomata* presents mostly tubular or flattened well-calcified zoaria (Figure 6.9); the apertures through which zooids are extruded are mostly of circular shape. Taxa of this order occur in the Ordovician-Holocene stratigraphical interval. *Order Trepostomata* of the Ordovician-Triassic stratigraphical interval frequently evolved massive, lamellate or stem-like colonies that contributed to the reef formation

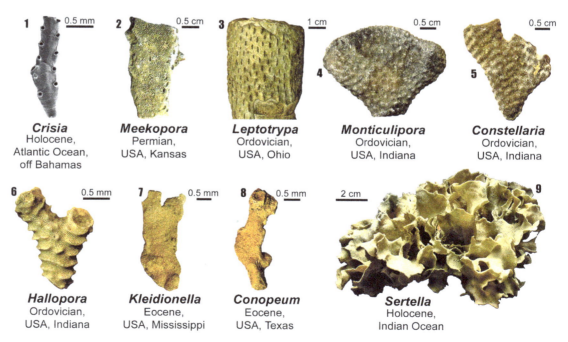

Figure 6.9 Examples of bryozoans of the classes Stenolemata (1–6) and Gymnolemata (7–9). 1–8: Specimens from the paleontological collections of the University of Calgary. 9: Specimen from the Senckenberg Natural History Museum, Frankfurt; published with permission.

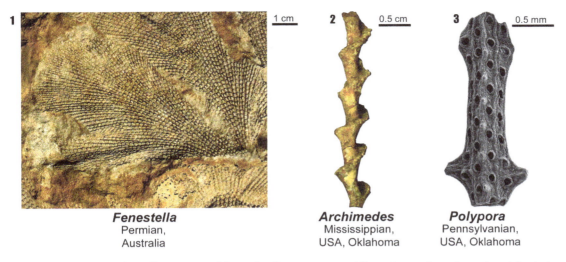

Figure 6.10 Examples of bryozoans of the order Cryptostomata. All specimens from the paleontological collections of the University of Calgary.

during different stratigraphical intervals (Figure 6.9); the zooids are often differentiated morphologically and with various functions in the colony. *Order Cryptostomata* is important not only for the contributions to the reef-building but also for the fact that some of the species of the order are excellent index fossils (e.g., *Archimedes*) (Figure 6.10); cryptostomates have a net-like appearance with the zooids aligned on the long branches of the colony. This group occurs in the Ordovician-Triassic stratigraphical interval.

Class Gymnolemata consists of taxa with encrusting or attached zoaria; zooids are often box-like and morphologically differentiated in the colony. Gymnolemates occur in the Ordovician-Holocene stratigraphical interval; they are frequent in the Paleozoic and abundant in the Late Mesozoic (Cretaceous) and Cenozoic. Post-Jurassic diversification of gymnolemates can be correlated with the decline of the representatives of the Class Stenolemata. The most diverse order of gymnolemates is *Cheilostomata* (Figure 6.9), which evolved in the Jurassic.

6.4 Brachiopods

Phylum Brachiopoda includes solitary taxa, some that can live in larger groups forming clusters, but never form colonies. The phylum evolved in the Early Cambrian and occurs in the fossil record throughout the Cambrian-Holocene stratigraphical interval; brachiopods are particularly frequent in the Paleozoic, but decline in diversity after the major crisis in the history of life from the proximity of the Permian/Triassic boundary. They are exclusively marine organisms and the highest diversity in the modern biotope is known from shallow waters with a depth of ~100 m; some brachiopod taxa can live to down to a depth of several thousand meters. The vast majority of brachiopods are benthic with only a small number of genera that evolved a planktic mode of life in the Cretaceous.

Brachiopods are extremely diverse, with more than 3600 genera known in living communities and fossil record; several major orders of brachiopods are known, most of them being exclusively Paleozoic, and the remaining ones—with one exception—were strongly reduced numerically at the Paleozoic/Mesozoic boundary (Figure 6.11). Such a good quality fossil record is due to the protective

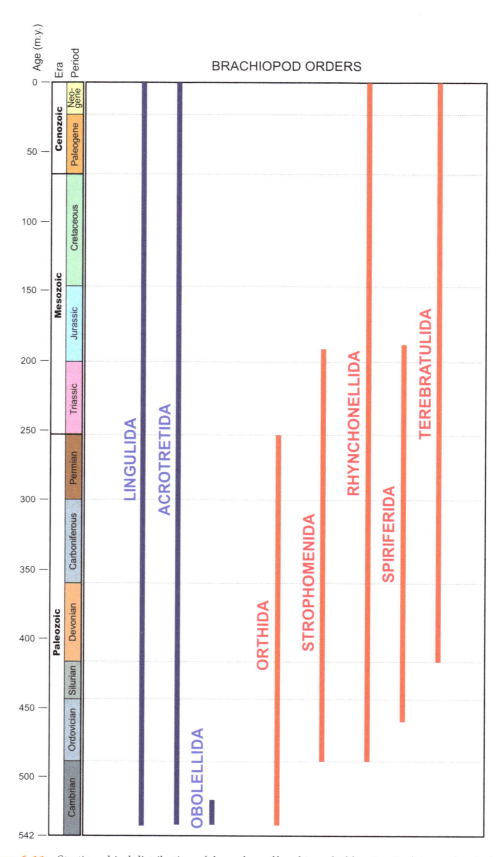

Figure 6.11 Stratigraphical distribution of the orders of brachiopods; blue-inarticulates, red-articulates.

exoskeleton consisting of two unequal valves of mineral nature that can fossilize relatively easily. The larger valve is also termed *pedicle valve* and bears a small opening known as *foramen* or pedicle opening through which a pedicle used for attachment to the substratum is extruded. Brachiopod shell part with the pedicle opening is the posterior part and the opposed one is termed frontal. The smaller valve is also termed *brachial valve* for it bears at the interior the lophophore support or *brachidium*. The two valves come in contact along the *commissure line*, which is subdivided into segments: posterior, anterior or frontal and two laterals. Most of the brachiopod taxa are bilaterally symmetrical and this feature can be seen in the shell morphology; the plane of symmetry is along the anterior-posterior direction and passes through the pedicle and middle of the frontal commissure. The bilateral symmetry is lost in many taxa adapted for reef environments; in such species and genera one of the valves evolved a conical shape, whereas the other one is reduced to a lid-like structure. The two valves are kept together with the aid of a ligament of organic and a hinge consisting of teeth and sockets, which are developed in the posterior part of the valves along the commissure line; the portion of the commissure line that bears the hinge is referred to as *hinge line, cardinal line*, or *cardinal commissure*.

The modern classification of the phylum Brachiopoda takes in consideration a variety of morphological features. Among the most significant ones are the shell chemical composition, presence/absence of a hinge, brachidium morphology and valve ornamentation and ultrastructure. Two classes of brachiopods are recognized: Inarticulata and Articulata; each of them consists of several orders; there are different and often divergent perspectives among specialists on how the brachiopod orders should be defined.

Class Inarticulata includes mostly taxa with chitinophosphatic shells (e.g., *orders Lingulida* and *Acrotretida*); among them the lingulids can be recognized by having in general antero-posteriorly elongate shells, whereas obollelids present mostly shells with a circular outline. As fossils, many of the representatives of this class have a distinct dark color with waxy appearance due to the organic matter in the valve composition (Figure 6.12). Valve surface is simple but some taxa evolved a simple ornamentation; valve interior presents attachment muscle scars and occasionally low-relief structures. Inarticulates represent the dominant brachiopod group in the Cambrian. One group of inarticulates namely *suborder Craniida* of the order Acrotretida is particularly important for they evolved calcitic shells; craniids have well-developed and characteristic muscle scars at the valve interior (Figure 6.12).

Class Articulata includes brachiopods with calcitic shell in which the two valves are kept together at the posterior end by a combination of a ligament and a hinge. The separation of this class into orders is largely based on the brachidium morphology, presence/absence of valve porosity, and the features of the posterior portion of the two valves. Five orders of the class Articulata are presented and in stratigraphical order of evolutionary occurrence they are: Orthida, Strophomenida, Rhynchonellida, Spiriferida, and Terebratulida.

Order Orthida (Cambrian-Permian) stratigraphical interval evolved early in the brachiopod evolutionary history being the earliest order of articulate brachiopods and became extinct at the major crisis of life at the Permian/Triassic boundary. In general they are small-sized biconvex species that have the shells ornamented with longitudinal costae (Figure 6.13). The hinge consists of small-sized teeth and

126   Chapter 6   Invertebrate Evolutions

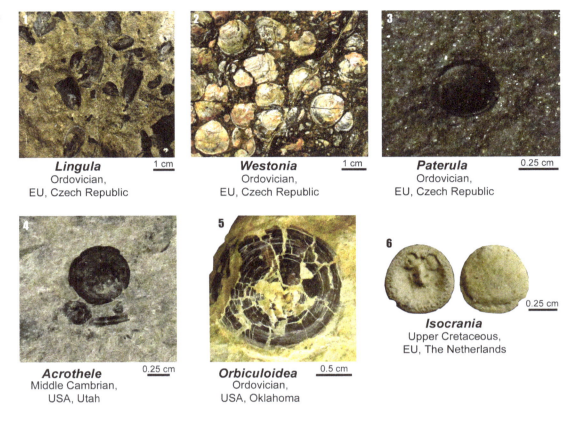

Figure 6.12 Examples of brachiopods of the class Inarticulata. All specimens from the paleontological collections of the University of Calgary.

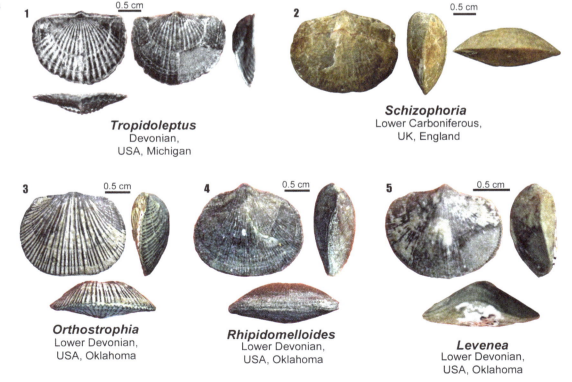

Figure 6.13 Examples of brachiopods of the order Orthida. All specimens from the paleontological collections of the University of Calgary.

sockets. Brachidium is absent or extremely simple consisting of short calcitic lamellae capable of supporting the lophophore base. The representatives of this order are considered in general *impunctate* but some morphologically advanced taxa evolved valve perforations. This order is one of the dominants in the brachiopod assemblages during the Ordovician times.

Order Stromphomenida (Ordovician-Lower Jurassic) is morphologically more diverse when compared with the representatives of the order Orthida and is subdivided into several suborders (Figure 6.14). The cardinal lines are in general long and straight and the two valves present different convexities. Valves are *pseudopunctate*, having pore-like structures that do not have openings on both surfaces of the valve. The main stock of this order is represented by *suborder Strophomenina* (Ordovician-Triassic). In this group the valve surface is ornamented with fine costae and well-developed growth lines; hinges are reduced. The representatives of this suborder were among the dominants in the brachiopod assemblages of the Ordovician-Devonian stratigraphical interval. *Suborder Productida* (Devonian-Permian) includes some large-sized taxa adapted for life in reefal environments; some most spectacular taxa of the reefal environments lost completely the valve symmetry (Figure 6.14). Productids

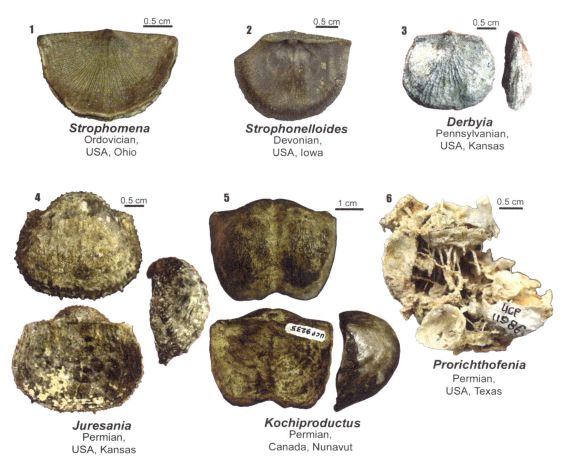

Figure 6.14 Examples of brachiopods of the order Strophomenida. All specimens from the paleontological collections of the University of Calgary.

were one of the dominants in the brachiopod assemblages of Late Carboniferous-Permian age.

Order Rhynchonellida (Ordovician-Holocene) is the brachiopod order with the longest evolutionary history and one of the dominant orders in the Mesozoic and Cenozoic times. Shells of the representatives of this order are small-sized and the valves present a variable convexity (Figure 6.15). Valve surface is ornamented with longitudinal ribs, which in rare taxa are reduced to very few conferring the shell a smoother aspect. The surface of the two valves is plicated, one feature that can be best seen in frontal view, where the commissure has a folded aspect. Most of the rhynchonellid taxa are impunctate. Hinge presents in general well-developed teeth and sockets. Brachidium consists of two blade-like calcitic structures termed *crura*.

Order Spiriferida (Middle Ordovician-Lower Jurassic) is morphologically diverse and for this reason is subdivided into several suborders. In the representatives of this order the brachidium evolved in two spirally coiled calcitic lamellae termed *spiralia*, each with conical aspect. The main stock of Spiriferida is represented by *suborder Spiriferina* (Silurian-Lower Jurassic). This suborder includes biconvex, occasionally large-sized shells with impunctate valves; the valve external surface is ornamented with fine longitudinal ribs and distinctly

Figure 6.15 Examples of brachiopods of the order Rhynchonellida. All specimens from the paleontological collections of the University of Calgary.

Figure 6.16 Examples of brachiopods of the order Spiriferida, suborder Spiriferina. All specimens from the paleontological collections of the University of Calgary.

plicate, one feature which is reflected in the folded frontal commissure and deflected and often folded lateral commissures. Spiralia present the cone apexes oriented toward the lateral sides of the shell (Figure 6.16). *Suborder Atrypidina* (Middle Ordovician-Devonian) is the oldest suborder of Spiriferida and most likely ancestral to the other three. Shells are biconvex and finely costate (Figure 6.17) with the spiralia oriented with the apexes toward the brachial valve or lateral sides. *Suborder Retziacea* (Silurian-Permian) includes punctate shells with general external features resembling those of rhynchonellid taxa (Figure 6.17); brachidium consists of spiralia with the apexes oriented toward the lateral sides. *Suborder Athyrididina* (Upper Ordovician-Lower Jurassic) is the only suborder of Spiriferida that survived the major cris is in the history of life at the Permian/Triassic boundary. Athyrididin shells have a general terebratulid aspect (Figure 6.17) with the brachidium consisting of laterally oriented spiralia. Different suborders of spiriferids were among the dominants of the brachiopod assemblages in the Devonian-Lower Carboniferous stratigraphical interval.

Order Terebratulida (Devonian-Holocene) is one of the two orders of living brachiopods together with the rhynchonellids and also one of the two dominant orders in the Mesozoic and Cenozoic times. Cardinal line is subangular under the foramen (Figure 6.18); brachidium is looped. The dominant group of terebratulids

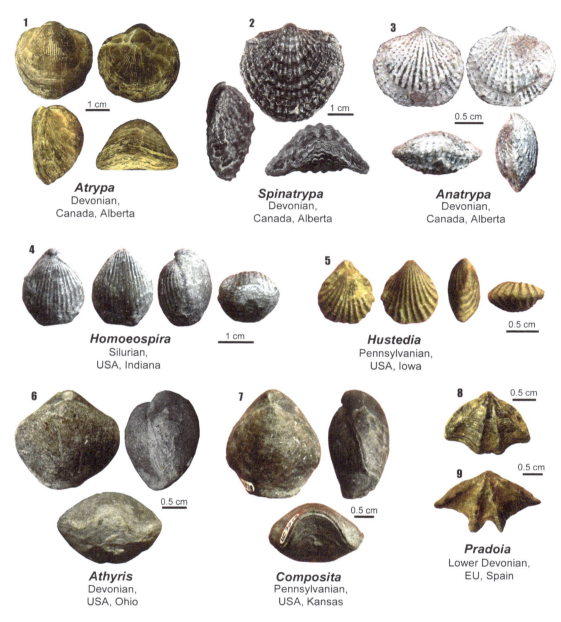

Figure 6.17 Examples of brachiopods of the order Spiriferida, suborders Atrypidina, Retziacea, and Athyrididina. All specimens from the paleontological collections of the University of Calgary.

is *suborder Terebratulidina* (Devonian-Holocene). Most of the taxa are small-sized but occasionally evolved larger shells such as *Stringocephalus* of the Devonian. The valves are in general smooth, without ornamentation but frequently with growth lines; valve surface is plicate especially in the Mesozoic representatives of this suborder. In the Cretaceous, *family Pygopidae* evolved a planktic mode of life. *Suborder Terebratellidina* (Triassic-Holocene) presents in general longer brachidia than the tererbratulidins; valve surface can be costate and with well-developed growth lines.

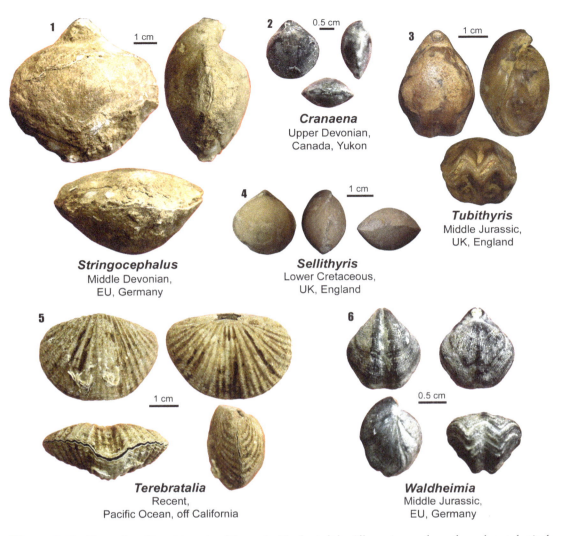

Figure 6.18 Examples of brachiopods of the order Terebratulida. All specimens from the paleontological collections of the University of Calgary.

6.5 Monoplacophorans, Polyplacophorans, and Scaphopods

These three classes of molluscs rise many questions about their origins and evolution and this is partly due to the sparse fossil record. None of them includes the earliest mollusc taxa of the Tommotian fauna; notably these small-sized taxa were included into two distinct classes solely based on the shell morphological features in the absence of data on the soft body.

Class Monoplacophora (Cambrian-Lower Devonian, Holocene) includes exclusively marine organisms that protect the soft body with a shell consisting of one calcitic piece. The shell is conical, straight or gently curved toward the apex, in other taxa flattened and bilaterally symmetrical, more rarely asymmetrical. Surface is smooth or ornamented with radial fine costae, growth lines or irregular structures. Most of the monoplacophoran genera occur as fossils from the Cambrian-Lower Devonian, but more frequently from the lower portion of this stratigraphical range. There is a gap that separates these occurrences from the Holocene ones; the latter include living taxa, such as *Neopilina*. This genus is of particular importance for it documents the occurrence of body segmentation in one mollusc group, raising the question if the molluscan ancestor was not an organism with body segmentation. It is questionable if the organisms known from Cambrian-Lower Devonian and Holocene occurrences belong to one single group or they represent two distinct groups with resembling hard body parts that evolved independently from different ancestors; in the latter case the resemblances in shell morphology can be regarded as the result of iterative evolution.

Class Polyplacophora is known mostly by the living representatives that include chitons (Figure 6.19) which live especially in intertidal conditions on rocky substratum, but can be occasionally found to a water depth of several thousand meters. Polyplacophorans are exclusively marine organisms. The soft body of the polyplacophoran organism is protected by a shell consisting of eight, more rarely seven plates that overlap each other in various degree; the calcareous plates of the shell can be fossilized and occur in general as isolated plates and more rarely in anatomical connection. The representatives of this class occur in the fossil record throughout the Upper Cambrian-Holocene stratigraphical interval but are rare.

Class Scaphopoda consists of bilaterally symmetrical taxa that have the soft body protected by an elongate calcareous shell opened at both ends. Scaphopods occur

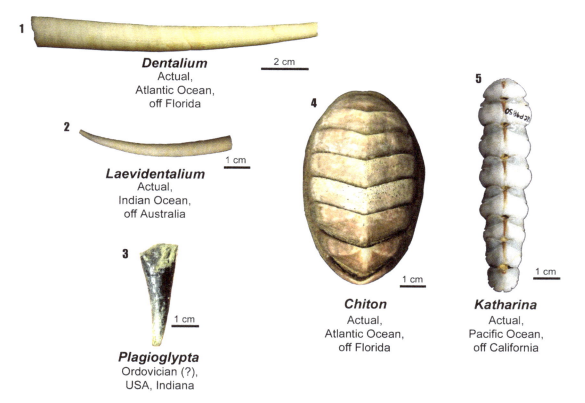

Figure 6.19 Examples of molluscs of the classes Scaphopoda (1–3) and Polyplacophora (4–5). All specimens from the paleontological collections of the University of Calgary.

in the fossil record in the Ordovician-Holocene stratigraphical interval; the fossil record of this class is sparse, with considerable gaps between different occurrences and partly for this reason their evolutionary history is poorly understood. The morphological features of the shell that can be used in evolutionary study are relatively reduced; most of them pertain of the simple shell gross architecture and ornamentation that can be costate, smooth or with growth lines. One interesting genus is *Plagioglypta* that presents circular grooves in the anterior and external portion of the shell, conferring it a multichambered appearance.

6.6 Gastropods

The representatives of this class, which are also known by the informal names of snails and slugs, have the soft body protected by an external calcareous shell consisting of only one piece, which is primarily coiled and presents a unitary interior, without partitions. Class Gastropoda evolved in the Early Cambrian times and probably the oldest known genus is *Aldanella*, a small-sized cosmopolitan taxon. The group diversified rapidly and adapted to all types of aquatic environments: fresh, brackish and marine waters; however, they were primarily marine organisms. Soft body is unsegmented, with a well-defined head and visceral mass presenting a distinct torsion; much of the soft body is formed by the foot, a muscular organ used for movement.

The general morphology of a gastropod shell shows that it can be coiled either dextrally or sinistrally. Most of the taxa present a trochospiral shell but planispiral coiling can occur in certain taxa; a distinct uncoiling pattern occurs in a relatively small number of genera and species. The end from which the shells starts to grow is the posterior one and corresponds to the apex, or the tip of the cone-like shell; the opposing end in the direction of growth that bears a large opening termed aperture is the anterior part of the shell. The shell consists of a variable number of whorls and two adjacent whorls form a depressed line termed suture. Anteriorly there is the *umbilicus*, a depression area in the proximity of the aperture. In many taxa there is an additional calcareous structure, namely the operculum that is used to close and seal the aperture in case the soft body is retracted within the shell.

Class Gastropoda is subdivided into three subclasses: Prosobranchia, Opistobranchia, and Pulmonata. This classification is based on the features that occur only in the soft body and do not include data from the shell morphology. Subclass Prosobranchia includes most of the known gastropod taxa and they are characterized by a complete torsion of the viscera. Subclass Opistobranchia consists of gastropods that lost the visceral torsion partially or completely. Subclass Pulmonata includes taxa adapted to terrestrial environments, in which the gills are replaced by a lung-like cavity of the mantle, which is heavily vascularized, which allows the organism to breath atmospheric oxygen.

Subclass Prosobranchia (Cambrian-Holocene) includes most of the known taxa of gastropods, living and fossils; in these taxa the mantle cavity is in anterior position and the aperture can be supplemented by an operculum. Most of the taxa of this class are marine, and only a smaller number of genera and species adapted to brackish and fresh water conditions. Two orders are recognized according to the gill morphology: Archaeogastropoda and Neogastropoda. *Order Archaeogastropoda* (Cambrian-Holocene) consists of taxa with *bipectinate gills*; the representatives of this order are particularly diverse and frequent in the Paleozoic (Figure 6.20). Order

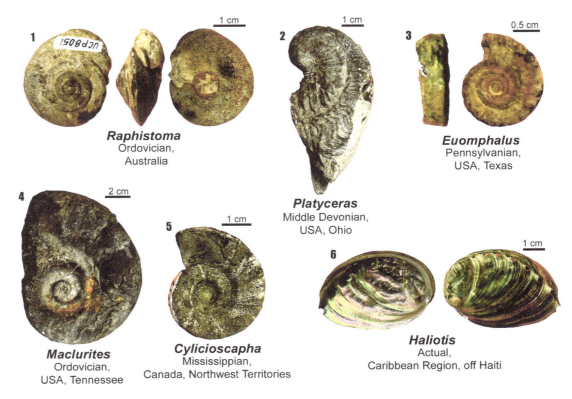

Figure 6.20 Examples of gastropods of the subclass Prosobranchia, order Archaeogastropoda. All specimens from the paleontological collections of the University of Calgary.

Neogastropoda (Ordovician-Holocene) includes gastropod taxa with *monopectinate gills* and is subdivided by many specialists into Mesogastropoda (Ordovician-Holocene) and Neogastropoda (Cretaceous-Holocene), but for pedagogical reasons a more conservative classification is presented herein. Neogastropods are diverse and abundant during the Mesozoic and Cenozoic times; most of the gastropod marine species in the modern seas and oceans belong to the neogastropod group (Figure 6.21).

Subclass Opistobranchia (Carboniferous-Holocene) includes taxa in which the calcareous shell is reduced and included within the mantle, or completely lost; they are mostly marine, rarely fresh water taxa. One interesting group of opistobranchs is represented by the *pteropods*, small-sized taxa that evolved a nektic mode of life; pteropods evolved in the Late Cretaceous and diversified in the Cenozoic.

Subclass Pulmonata (Carboniferous-Holocene) consists of gastropods that adapted successfully to the terrestrial environments, and secondarily returned to aquatic (fresh water) conditions (Figure 6.22). Such colonization of land by gastropods was supported by the evolution of the capability to breathing atmospheric oxygen.

Gastropod higher units are defined according to a relatively small number of morphological features, which are provided by the soft body morphology as already mentioned. Such units have small chances to provide a coherent perspective on the group's evolution from the perspective of the fossil record. The maximum information yielded by such classification is that the recognized groups have different stratigraphical ranges and some groups are older than the others. For this reason reliable evolutionary studies on the representatives of class Gastropoda are realized at lower taxonomic levels, such as species and genus.

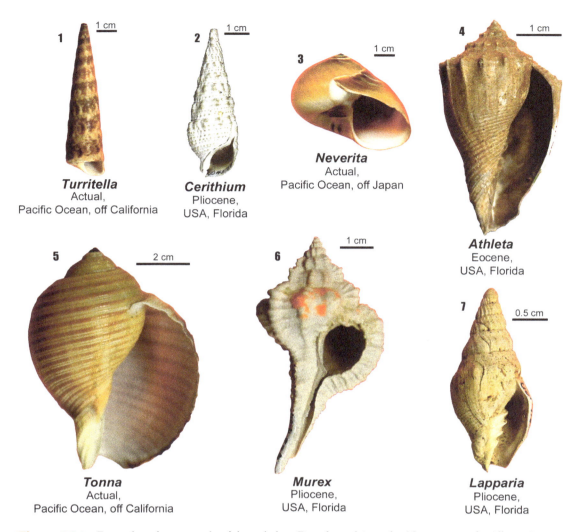

Figure 6.21 Examples of gastropods of the subclass Prosobranchia, order Neogastropoda. All specimens from the paleontological collections of the University of Calgary.

Figure 6.22 Example of gastropod of the subclass Pulmonata. Specimen from the paleontological collections of the University of Calgary.

6.7 Bivalves

Class Bivalvia includes molluscs without a head and with the soft body covered and protected by two calcareous valves. The two valves are equally developed only in a small number of genera and species, and are kept together by a ligament of organic nature and a hinge consisting of a combination of teeth and sockets complimentary developed on the two valves; therefore, one tooth on one valve corresponds to one socket on the other. The valve interior presents several features left by the different organs of the soft body, such as *pallial line* that indicates where the mantle was attached to the valve, *pallial sinus* that marks the position of the siphon and *muscle scars* that mark the position of the muscles used to open and close the valves. Bivalves are bilaterally symmetrical organisms and the plane of symmetry passes between the two valves; however, the concept of symmetry should be considered in a rather loose sense for the two valves are not equal in most of the taxa. Symmetry is lost especially in the case of taxa adapted to reefal environments.

All bivalves are taxa adapted to aquatic environments; primarily they were marine organisms, which in the course of the group's evolution colonized brackish and fresh water environments. They are benthic organisms living on the sea floor or digging in the superficial portion of the sediments. In general bivalves have reduced capabilities of movement; movement usually happens with the aid of a muscular organ termed foot, but some species developed a different system realized by creating a current of water through the mantle.

Bivalve classification is based on a variety of morphological features of the two valves and among them the hinge or dentition is paramount. Several subclasses are recognized within the bivalves and five of them are presented herein: Palaeotaxodonta, Pteriomorphia, Palaeoheterodonta, Heterodonta, and Anomalodesmata. They are not evolutionary units but most likely polyphyletic units in which a certain feature or groups of features were achieved through convergent and iterative evolution.

Subclass Palaeotaxodonta (Cambrian-Holocene) consists of taxa that mostly have aragonitic valves. The valves are equal in size and with the hinge consisting of a multitude of small-sized teeth that can be straight or curved and oriented perpendicularly or at a high angle to the valve margin; this type of dentition is referred to as *taxodont*. The taxa of this subclass are rarely encountered in the assemblages of living bivalves. Genus *Nucula* is a typical representative of this subclass (Figure 6.23), which includes only the *order Nuculoida*.

Subclass Pteriomorphia (Cambrian-Holocene) is a diverse group of marine bivalves that consists of three orders: Arcoida, Mytiloida, and Pterioida. *Order Arcoida* includes taxa with taxodont dentition and valve surface ornamented with fine costae; genera such as *Arca, Anadara, Cucullaea,* and *Glycimeris* are included within order Arcoida (Figure 6.23). Order Mytiloida consists of taxa with elongate shells, equal valves and dentition presenting one elongate tooth parallel to the valve margin; such type of dentition is known as *dysodont*. Two examples of genera included in this order are *Mytilus* and *Pinna* (Figure 6.24). Order Pterioida is morphologically diverse and include taxa with equal or unequal valves, and different dentition types. Genera such as *Pecten, Euvola, Crassosterea, Gryphaea,* and *Inoceramus* are included in this order (Figure 6.24). Some of them are used extensively in biostratigraphy and a classical example is that of *Inoceramus* and allied genera. *Inoceramids* occurred in the Permian, are most frequently recorded in the Cretaceous and became extinct at the Cretaceous/Paleogene boundary. Morphologically the valves present prominent

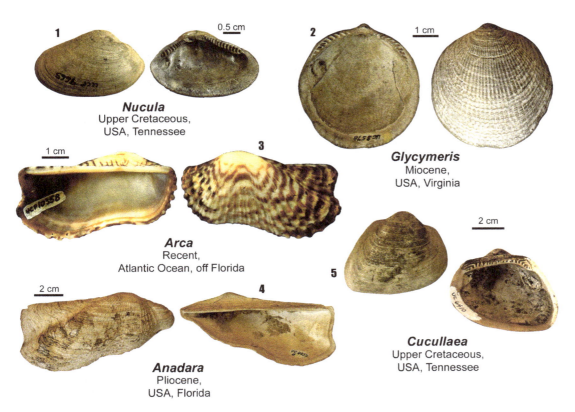

Figure 6.23 Examples of bivalves of the subclass Palaeotaxodonta (1) and subclass Pteriomorphia, order Arcoida (2–5). All specimens from the paleontological collections of the University of Calgary.

concentrical growth lines; other ornamentation elements, such as tubercles, costae, etc., are rare. Valves are very thin and for this reason rarely preserved; the calcitic valves are usually disintegrated into the component calcite crystals that can be found in the micropaleontological samples as *inoceramid prisms*. Inoceramids are often large sized and the largest valves can be up to 1 m in length. This group of marine bivalves is frequent in the shelf and basin environments.

Subclass Palaeoheterodonta (Ordovician-Holocene) includes taxa with equal or unequal valves and diverse dentitions. Most of the genera included in this subclass are Paleozoic where they occur mostly as minor components of the bivalve assemblages. Primarily they were marine taxa, but colonized the brackish and fresh water environments in the Mesozoic and Cenozoic. Three orders are included within this subclass: Modiomorphoida, Unionoida, and Trigonoida. *Order Modiomorphoida* includes taxa with primitive morphological features and for this reason considered ancestral to many of the Mesozoic and Cenozoic bivalve groups; *Modiomorpha* is an example of genus of this order (Figure 6.25). *Order Unionoida* evolved after the Permian/Triassic boundary. They evolved during Mesozoic and achieved the maximum diversity in the Cenozoic. Unionoids are adapted to the fresh water and brackish water environments. They present valves of nearly equal size and a wide variability in the hinge development. Valve surface frequently presents growth lines, costae and tubercles; *Unio* is an example of genus of this order (Figure 6.25). *Order Trigonoida* evolved in the Devonian and diversified rapidly during the Mesozoic, especially in the Jurassic and Cretaceous. Trigonoids have a rhomboidal outline and truncated valves; the two valves are smooth or present strong ornamentation consisting of

138   Chapter 6   Invertebrate Evolutions

Figure 6.24 Examples of bivalves of the subclass Pteriomorphia, order Mytiloida (1–2) and order Pterioida (3–6). All specimens from the paleontological collections of the University of Calgary.

Figure 6.25 Examples of bivalves of the subclass Palaeoheterodonta. All specimens from the paleontological collections of the University of Calgary.

costae with various orientation and tubercles. *Trigonia* and *Linotrigonia* are examples of genera of this order (Figure 6.25).

Subclass Heterodonta (Ordovician-Holocene) includes most of the marine taxa of the living bivalve assemblages. Three orders are included in this subclass: Veneroida, Myoida, and Hippuritoida. *Order Veneroida* is diverse and dominant in most of the living bivalve assemblages; the largest living bivalve genus is included in this order: *Mactra*. Other veneroid genera are *Venus*, *Chama*, *Dosinia*, *Pitar*, etc. (Figure 6.26). *Order Myoida* includes taxa which are burrowers and often present highly modified dentition; one example is the genus *Cyrtopleura* that evolved a spectacular ligament support termed *chondrophore* (Figure 6.27). *Order Hippuritoida* representatives are often referred to as *rudist bivalves* and evolved in the Late Jurassic. Mesozoic representatives of this group are frequent major reef-builders (Figures 6.27 and 6.28). Hippuritoids are bivalves with aberrant morphology in which the valves and dentition are asymmetrically developed. One valve is cone-shaped and the other reduced, lid-like in the case of the genera permanently attached to the sea floor (e.g., *Praeradiolites*).

Subclass Anomalodesmata (Ordovician-Holocene) consists of marine taxa with mostly aragonitic valves. Living representatives of this subclass occur frequently in deep oceanic environments. *Pleuromya* is an example of genus of this subclass, which includes only the *order Pholadomyoida* (Figure 6.27).

Figure 6.26 Examples of bivalves of the subclass Heterodonta, order Veneroida. All specimens from the paleontological collections of the University of Calgary.

Figure 6.27 Examples of bivalves of the subclass Heterodonta, order Myoida (1), order Hippuritoida (3), and subclass Anomalodesmata (2). All specimens from the paleontological collections of the University of Calgary.

Figure 6.28 Example of rudist bank with the specimens fossilized in life position. Specimen from the paleontological collections of the University of Calgary.

6.8 Nautiloid Cephalopods and Closely Related Major Groups

Class Cephalopoda is subdivided into six subclasses: Nautiloidea, Endoceratoidea, Actinoceratoidea, Bactritoidea, Ammonoidea, and Coleoidea. Living cephalopods are represented by about 700 species, but the fossil record consists of a number at least 20 times higher. The known evolutionary history of the *class Cephalopoda* began in the Late Cambrian times and by the Early Ordovician is documented the earliest diversification, which resulted in the evolution of several shell architectural types. Cephalopod shell is external in most of the representatives of the class; in only one subdivision it has an internal position (endoskeleton) and in some living representatives of this subclass it is strongly reduced or completely lost. The shell consists of one piece, is mostly aragonitic and presents two general morphologies: one for the taxa in which is external and one for those in which the shell is transformed in an endoskeleton. In the former case it is multichambered and the living organism occupies only the last-formed chamber, whereas the older ones remain empty; the empty chambers are in connection through the siphon or siphuncle that can be in central or eccentric position; two successive chambers are separated by a septum, which is of mineral nature. Other fossilized structures from cephalopods are paired opercula or *aptychi* and calcified portions of the jaws, which are known under the general term of *rhyncholites*. Modern cephalopods are exclusively marine organisms that can develop their life cycles from the nearshore conditions to the deepest portions of the oceans. Fresh or brackish water cephalopods are not known and there is no indication from the fossil record of this group that they ever colonized environments characterized by lower than marine salinity. Cephalopods are carnivorous and predatory organisms.

Subclass Nautiloidea is the oldest subclass of cephalopods; its stratigraphical range is Upper Cambrian-Holocene. The earliest representatives of this subclass were small-sized and with a gently curved shell axis. One of the most diverse orders of this subclass, namely *order Orthocerida* evolved in the Early Ordovician times and diversified quickly afterwards. Orthoceratids present in general straight (e.g., *Orthoceras*, *Michelinoceras*, etc.) or gently curved shells (e.g., *Peismoceras*, *Adelphoceras*, etc.) and there is a distinct evolutionary trend toward shell coiling in the Silurian-Devonian representatives of this order (Figure 6.29). Notably, the coiling is loose and in general the successively formed whorls are not in contact. Orthoceratids became extinct in the Triassic. *Order Oncocerida* (Late

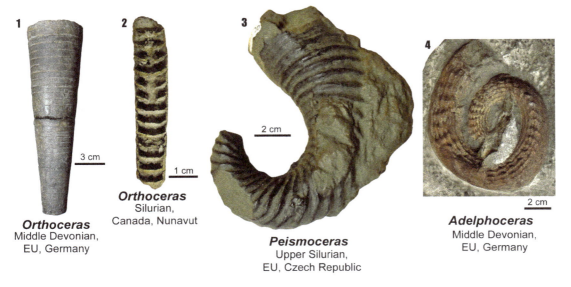

Figure 6.29 Examples of cephalopods of the subclass Nautiloidea, order Orthocerida. All specimens from the paleontological collections of the University of Calgary.

Figure 6.30 Examples of cephalopods of the subclass Nautiloidea, orders Discosorida (1) and Oncocerida. Both specimens from the collections of the Museum of Natural History, Berlin; published with permission.

Ordovician-Early Carboniferous) consists of taxa with short growth axis and wider shells (e.g., *Oncoceras*, etc.) (Figure 6.30); they are of particular importance from an evolutionary perspective for they represent the ancestors of the nautiloid group that includes the living genus *Nautilus*. *Order Discosorida* (Middle Ordovician-Devonian) is a nautiloid order that probably evolved from the most primitive representatives of this subclass (e.g., *Phragmoceras*, etc.) (Figure 6.30). The typical representatives of this subclass are included within the *order Nautilida* (Devonian-Holocene); this is the only order of nautiloids with living representatives (Figure 6.31). Nautilids have tightly coiled shells with simple suture lines that mark the junction between septa and shell wall; siphuncles occupy a central position. *Nautilus*, *Proclydonautilus*, *Clydonautilus*, and *Eutrephoceras* are examples of genera of the order Nautilida.

Subclass Endoceratoidea evolved from orthoceratoid ancestors and includes large (up to 4 m in length) straight-shelled cephalopods. Siphuncle is wide, eccentric in position and partly filled with calcitic cone-in-cone structures used to control the shell position. Endoceratoids were major predators in Ordovician and Early Silurian.

Subclass Actinoceratoidea also evolved from orthoceratoid ancestor and presents large-sized shells with a wide siphuncle; shell position during organism life was controlled with the aid of siphuncle and cameral deposits of calcareous nature. Actinocetratoids evolved in Ordovician and became extinct in the Late Carboniferous.

Subclass Bactritoidea consists of taxa with small-sized shells, straight or slightly curved, a few centimeters in length (Figure 6.32). The eccentric siphuncle is marginal in position and this feature is of paramount importance in considering the bactritoid group the most probable ancestor of ammonoids and coileoids. The stratigraphical interval in which the representatives of subclass Bactritoidea occur is Silurian-Devonian. Genus *Bactrites* is a typical example of this subclass.

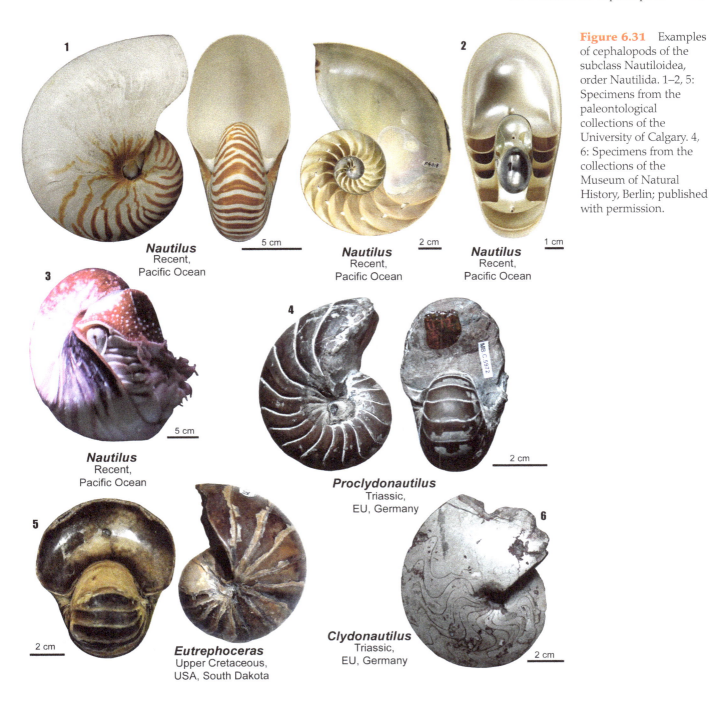

Figure 6.31 Examples of cephalopods of the subclass Nautiloidea, order Nautilida. 1–2, 5: Specimens from the paleontological collections of the University of Calgary. 4, 6: Specimens from the collections of the Museum of Natural History, Berlin; published with permission.

6.9 Ammonoid Cephalopods

The informal term *ammonoid* is used for the representatives of subclass Ammonoidea. They are organisms in which the shell consists of one piece and in the general aspect resembles the shell of the coiled nautiloids of the order Nautilida. A more detailed comparative study of the shell morphology of subclass Ammonoidea and order Nautilida indicates that the morphological differences are of considerable magnitude. Among them probably the most significant ones are the siphuncle position and sutural line morphology. Siphuncle is situated at the septum center in nautilids and occupies an eccentric position in ammonoids. Sutural line is straight or gently curved in nautilids, whereas in ammonoids is considerably more complex

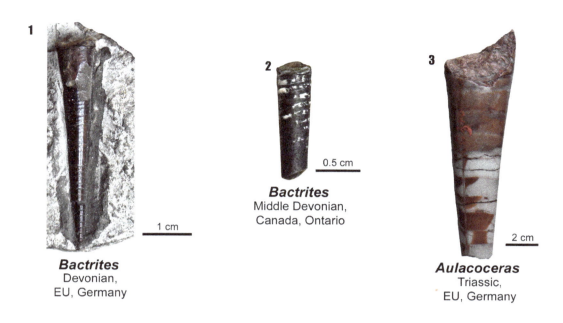

Figure 6.32 Examples of cephalopods of the subclass Bactritoidea (1–2) and Coleoidea (3). 1, 3: specimens from the collections of the Museum of Natural History, Berlin; published with permission. 2: specimen from the paleontological collections of the University of Calgary.

and there is gradual increase through time in the complexity of this feature. Subclass Ammonoidea evolved in the Devonian and became extinct at the Cretaceous/Paleogene boundary; in fact ammonites represent one classical example of a major group that became extinct at this major extinction event in the history of life.

Morphological features taken in consideration in ammonoid classification are extremely diverse, but they can be subdivided in several categories. *Coiling mode* refers to the pattern of addition if new portions of the shell: planispiral (spiral in one plane), trochospiral (three-dimensional coil), streptospiral (without a regular pattern), combinations of two or three patterns, etc.; the term *uncoil* refers to the development of at least one pattern different from that in the earliest stage and frequently of a straight portion following a coiled stage. *Sutural line* presents different morphologies that are often consistent at order level and for this reason they are often termed according to the order in which they dominate (e.g., goniatitic, ceratitic, ammonitic, etc.). *Chamber overlapping* qualitatively describes how the newly added portions of the shell (whorls) cover the previously formed ones. An evolute shell is that in which the last whorl does not cover the previously formed ones, which remain visible; if there is a complete overlapping the shell is termed involute; intermediary shells are termed evolute-involute or involute-evolute according to the degree of overlapping. *Ornamentation* refers to external morphological features such as keels, spines, tubercles, nodules.

The origins of ammonoids are well-documented. Erben (1966) demonstrated the occurrence of a transition between straight-shelled bactritoids and earliest coiled ammonoids of the *order Goniatitida* in the Devonian times; the representatives of this order became extinct at the Permian/Triassic major crisis in the history of life. Goniatites evolved a new type of sutural line that is known as goniatitic, which consists of an alternance of lobes and saddles that often have an angular or subangular aspect. Due to the fast peace of evolution and short stratigraphical ranges, some of the representatives of order Goniatitida are excellent index fossils in the Upper Paleozoic. Some of the representative genera of this order are *Goniatites*, *Manticoceras*, *Tornoceras*, and *Metalegoceras* (Figure 6.33). Other ammonoid orders evolved in the Paleozoic are *Anarcestida* (e.g., *Anarcestes*) (Figure 6.33) and *Clymeniida* that are restricted to the Devonian, and *Prolecanitida* that survived the Permian/Triassic boundary crisis and became extinct in the Triassic times.

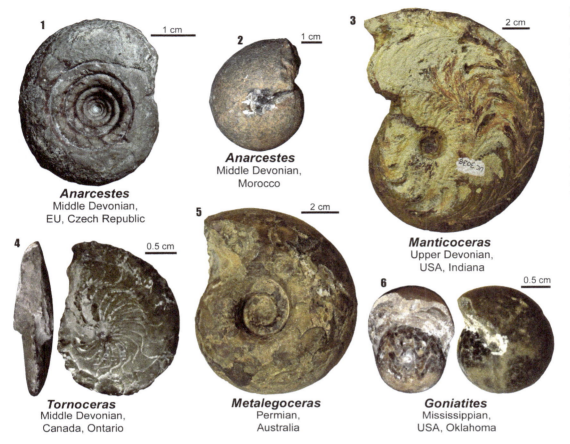

Figure 6.33 Examples of cephalopods of the subclass Ammonoidea, orders Anarcestida (1–2) and Goniatitida (3–6). 1: Specimen from the collections of the Museum of Natural History, Berlin; published with permission. 2–6: Specimens from the paleontological collections of the University of Calgary.

Order Ceratitida began their evolution in the late Paleozoic, survived the Permian/Triassic crisis and evolved rapidly in the Triassic; they are the most frequent ammonoids of the Triassic times. Ceratitic sutural line is the dominant type among the representatives of this group. Some ceratites evolved shells with prominent ornamentation but most taxa are smooth or lightly ornamented; all the taxa of this order are tightly coiled (Figure 6.34). The high rates of evolution together with the short stratigraphical ranges of many species make the representatives of order Ceratitida excellent index fossils for the Triassic. Ceratites did not survive the Triassic/Jurassic crisis. Examples of ceratitid genera: *Discoceratites*, *Joannites*, and *Dieneroceras*.

Order Phylloceratida of the Triassic-Cretaceous stratigraphical interval is particularly important from evolutionary perspective for it is the ancestor of the dominant orders of the Jurassic and Cretaceous. Phylloceratids are tightly coiled and in general with weak ornamentation. The representatives of this order occur as minor components of the ammonoid assemblages throughout their stratigraphical range. *Phylloceras* is the most frequent genus of this order (Figure 6.34).

Order Lytoceratida (Jurassic-Cretaceous) presents a wide range of morphological variability. Some taxa such as genus *Lytoceras* of the Jurassic-Cretaceous are tightly coiled and with evolute or evolute-involute shells; frequently the shells of lytocearids present constrictions, which can be used in certain cases as clues in identification (Figure 6.35). One of the most spectacular trends in the representatives of this order is the evolution of uncoiled shells during the Middle Jurassic-Cretaceous (e.g., *Spiroceras*, *Ancyloceras*, *Acanthoscaphites*, *Scalarites*, *Nipponites*, *Ainoceras*, *Eubostrycoceras*, etc.) (Figures 6.35 and 6.36). Such genera present two or more coiling patterns throughout ontogeny and *Nipponites* is completely irregularly coiled.

146 Chapter 6 Invertebrate Evolutions

Figure 6.34 Examples of cephalopods of the subclass Ammonoidea, orders Ceratitida (1–3) and Phylloceratida (4). 1: Specimen from the Senckenberg Natural History Museum, Frankfurt; published with permission. 2–4: Specimens from the paleontological collections of the University of Calgary.

Figure 6.35 Examples of cephalopods of the subclass Ammonoidea, order Lytoceratida. 1–2, 4–5: Specimens from the collections of the Museum of Natural History, Berlin; published with permission. 3, 6: Specimens from the paleontological collections of the University of Calgary.

Figure 6.36 Examples of cephalopods of the subclass Ammonoidea, order Lytoceratida. All specimens exhibited in the National Museum of Nature and Science, Tokyo; base photographs courtesy Dr. K. Tanaka; published with permission.

Order Ammonitida includes most of the ammonoids of the Jurassic-Cretaceous stratigraphical interval. The taxa of this order evolved rapidly and for this reason are extensively used in biostratigraphy. The vast majority of the ammonitids have tightly coiled shells that are often ornamented. *Dactylioceras, Amaltheus, Hildoceras, Harpoceras, Schloenbachia, Cadomites, Stephanoceras, Garantiana, Perisphinctes, Scarburgiceras, Ataxioceras, Polyptichites, Schlotheimia, Tissotia,* and *Placenticeras* are examples of representative genera of the order Ammonitida (Figures 6.37 and 6.38). Gigantism is encountered among some genera of the Latest Cretaceous (Campanian-Maastrichtian), when shells with a diameter of over 2 m are known.

Other ammonoid fossils are calcified jaws, which are known under the general term of rhyncholites. They are differentiated into *rhyncholites* that are calcified upper jaws and *conchorhynchs*, which are calcified lower jaws (Figure 6.39). The stratigraphical interval in which rhyncholites are encountered extends above the Cretaceous/Paleogene boundary where the ammonoids became extinct; therefore, it cannot be ruled out that some of them are calcified jaws of nautilids. *Aptychi* are operculum-type structures that evolved in some cephalopods, most likely ammonoids, to which they are traditionally assigned. Nautilid origins for a part of the aptychi should also be taken in consideration (Figure 6.39).

Figure 6.37 Examples of cephalopods of the subclass Ammonoidea, order Ammonitida. 1–5: Specimens from the collections of the Museum of Natural History, Berlin; published with permission. 6–9: Specimens from the paleontological collections of the University of Calgary.

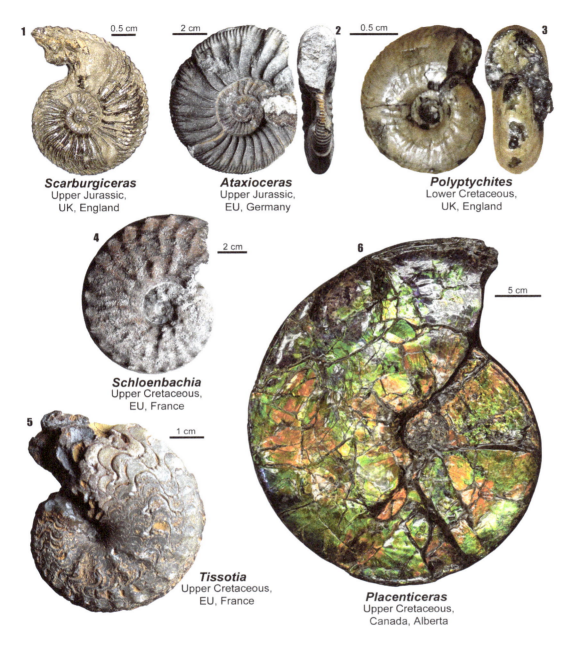

Figure 6.38 Examples of cephalopods of the subclass Ammonoidea, order Ammonitida. 1, 3, 6: Specimens from the paleontological collections of the University of Calgary. 2, 4–5: Specimens from the collections of the Museum of Natural History, Berlin; published with permission.

Figure 6.39 Examples of rhyncholites (1–2) and aptychi (3). 1–2: Specimens from the collections of the Museum of Natural History, Berlin; published with permission. 3: Specimen from the Museum of Natural Sciences, Paris.

6.10 Coleoid Cephalopods

Coleoid cephalopods are frequent occurrences in the modern seas and oceans especially in deeper water environments. By contrast to the nautiloid living genus *Nautilus* that has four gills, living coileoids have only two and for this are referred to as *dibranchiate* cephalopods. The representatives of this subclass underwent significant morphological changes during the long evolution of the group and it appears a rather difficult task to recognizing the modern coleoid body plan among the earliest taxa of Late Paleozoic age. The stratigraphical range of subclass Coleoidea is Devonian-Holocene. All the hard body parts of a coleoid animal are internal forming an endoskeleton, a setting that contrasts with all the other subclasses of cephalopods where the shell is external.

General morphology of the hard body parts of one coleoid shows the occurrence of three parts, which from the posterior toward anterior part are: rostrum, phragmocone and pro-ostracum. The septate portion of the coleoid shell is the aragonitic and thin-walled *phragmocone* that has the shape of a cone with anterior opening; this structure presents one thin siphuncle that marks the ventral side of the organism. The initial chamber of the shell, which is termed *protoconch*, is situated at the posterior end of the phragmocone. Anteriorly from the phragmocone is the flattened aragonitic *pro-ostracum*. Due to their aragonitic nature, phragmocone and pro-ostracum are rarely preserved. *Rostrum* is the segment of coleoid hard body parts that occurs frequently in the fossil record; this is because of its calcitic nature. During the life time of the coleoid organism the rostrum had a role in balancing the body position; this function is indicated by the high density of this structure, which consists of numerous layers of radial calcite microcrystals adjacent to each other. Rostrum, phragmocone and pro-ostracum are differently developed in various coleoid groups. Three orders of the subclass Coleoidea are presented herein: Belemnitida, Sepioida, and Octopoda.

Order Belemnitida (Devonian-Cretaceous) is the oldest of the subclass Coleoidea and the only recorded during the Devonian-Triassic stratigraphical interval. Rostrum is well-developed in the representatives of this order and most of the belemnite fossils are rostra; phragmocone and pro-ostracum occur rarely, whereas soft body parts only in cases of exceptional preservation. Genera *Belemnoteuthis*, *Plesioteuthis*, *Pachyteuthis*, *Hastites*, *Actinocamax*, and *Belemnitella* are examples of belemnites (Figure 6.40). *Megateuthis* is a Jurassic genus that could reach larger size and for this reason is considered a classic example of belemnite gigantism. Belemnites evolved rapidly in the Cretaceous times and a high-resolution biostratigraphical framework based on the representatives of this order is in use today. Less known is the fact the name of belemnites is the oldest given to a group of fossils and which is still in use today. More the 23 centuries ago the Creek philosopher Aristotle of Stagira gave the name *belemnos* (=arrowhead) to the fossil rostra.

Order Sepioida (Jurassic-Holocene) presents well-developed phragmocone and pro-ostracum that contrast to the rostrum that is considerably reduced when compared to those of the representatives of order Belemnitida; *Actinosepia* and *Spirulirostra* are examples of sepioid genera (Figure 6.41). The most advanced coiling is in the living genus *Spirula*, a deep-water genus, in which the phragmocone consists of about two whorls that are not adjacent (Figure 6.42). The eccentric siphuncle that is adjacent to the phragmocone margin is a strong argument in demonstrating the belemnoid origins of the sepioids.

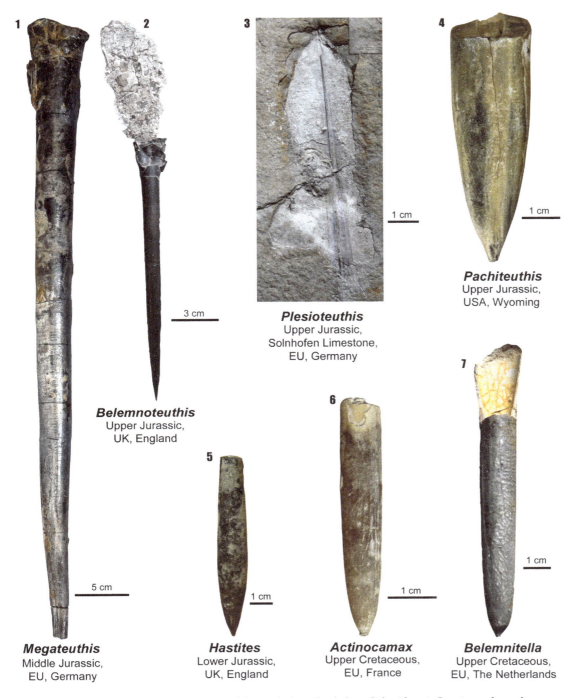

Figure 6.40 Mesozoic representatives of the cephalopod subclass Coleoidea. 1: Specimen from the Senckenberg Natural History Museum, Frankfurt. 2–3, 7: Specimens from the collections of the Museum of Natural History, Berlin; published with permission. 4–6: Specimens from the paleontological collections of the University of Calgary.

Order Octopoda (Cretaceous-Holocene) lost completely the hard body parts during the process of evolution. For this reason they are extremely rare in the fossil record. One rare octopod that evolved hard body parts is *Argonauta*, which is also known as the paper nautilus (Figure 6.24). The coiled and ornamented planispiral shell is used to protect the eggs until they hatch.

Figure 6.41 Examples of sepioid genera of the cephalopod subclass Coleoidea. 1: Specimen from the paleontological collections of the University of Calgary. 2: Specimen illustrated by Tomida (1992, fig. 4: 1); © Palaeontological Society of Japan; published with permission.

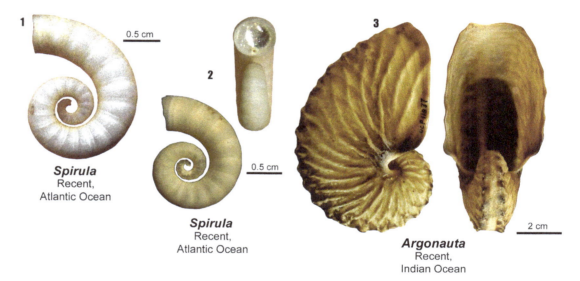

Figure 6.42 Examples of sepioid (1) and octopod (2) hard body parts of living taxa. All specimens from the paleontological collections of the University of Calgary.

6.11 Trilobites

Arthropods evolved rapidly after the "agronomic revolution" at the Proterozoic/Phanerozoic boundary and two of the most famous lagerstätten of the Cambrian, namely Chengjiang fauna of China (late Early Cambrian age) and Burgess Shale of Canada (Middle Cambrian age) yielded rich arthropod assemblages. These exceptionally preserved assemblages, by the diversity and complexity of the arthropod body planes indicate a longer evolution process that led to such achievements. The Atdabanian stage of the Lower Cambrian records a major event in the diversification of life on Earth and this is the occurrence and rapid diversification of trilobites; such process is well-documented due to the evolution of a protective chitinous carapace that was partly mineralized with up to 40% calcium carbonate. Trilobites are the oldest arthropods in the fossil record and fossils from below the Proterozoic/Phanerozoic boundary seemingly indicate that the group evolved from Ediacaran organisms (e.g., *Onega*, *Yorgia*, etc.); notably, a direct series of taxa leading from some ediacarans to trilobites is not well documented at this time.

Trilobites were mostly benthic marine organisms throughout their evolutionary history, with only one order evolving a planktic mode of life; there is no evidence do document if the representatives of this subphylum colonized fresh and brackish water environments. However, trace fossils indicate that some trilobite species could walk on short distances one the beaches. The fossil record of the group shows that shortly after the evolution in the Early Cambrian times (Atdabanian) the group underwent a rapid diversification that led to a peak in diversity in the Late Cambrian. Another diversification, this time at order level happened during the Ordovician. Throughout the Silurian-Devonian stratigraphical interval the trilobites were in continuous decline and ultimately became extinct shortly before the Permian/Triassic boundary. The representatives of this subphylum owe their success to several innovations some of which are mentioned here: evolution of the articulated protective carapace that provided increased protection against predators, development of numerous individuals of the same species that is a classic example of "safety in numbers," and evolution of a large pair of eyes that provided a wide visual field and helped in spotting the predators.

Trilobite carapace shows a remarkable homogeneity in chemical composition in one individual. Moreover, this structure extends providing protection to the appendages also. On the dorsal side, the cephalon consists of a central prominent part or *glabella* and the two eyes are symmetrically situated on its two sides. *Cephalon* lateral portions are separated into a proximal region that is immobile (*fixigena*) and a distal one that is mobile (*librigena*); the two are separated by the *facial suture* that has a paramount role in taxonomy and classification. Body segmentation is mostly apparent in the *thorax* where the axial and lateral zones are subdivided in segments termed pleura; each pleuron can terminate laterally with a spine. *Pygidium* consists of fused segments that in some taxa are not completely fused and sutures between them remain visible as shallow furrows. Spines can occur in different positions on the dorsal side, but the most important from a taxonomic perspective is the *caudal spine* situated at the posterior end of the pygidium.

At least eight orders are recognized in the modern classifications of trilobites. Shortly after group's evolutionary occurrence trilobites are represented by large-sized taxa. One such example is *Cambropallus* of the Early Cambrian that frequently presents a length of over 0.25 m (Figure 6.43); this genus belongs to the *order Redlichiida*, which is exclusively Cambrian. Other major taxonomic units restricted to the Cambrian are *order Olenellida* and *order Corynexochida* (Figure 6.44). Cambrian was the time of experimentation in trilobite evolution: the representatives of *order Agnostida* evolved a planktic mode of life in the Middle Cambrian times (Figure 6.43). Agnostid trilobites are small-sized, with a length of several millimeters, almost equal in size cephalon and pygidium, two to three thoracic segments and lack eyes completely. Notably, blind trilobite taxa also occur in the deep water environments in which the light could not penetrate. Only one trilobite order occurs in the fossil record above the Devonian/Carboniferous boundary: *order Proetida*. In the Permian times some of the representatives of this order evolved gigantic taxa, with a length of about 0.80 m.

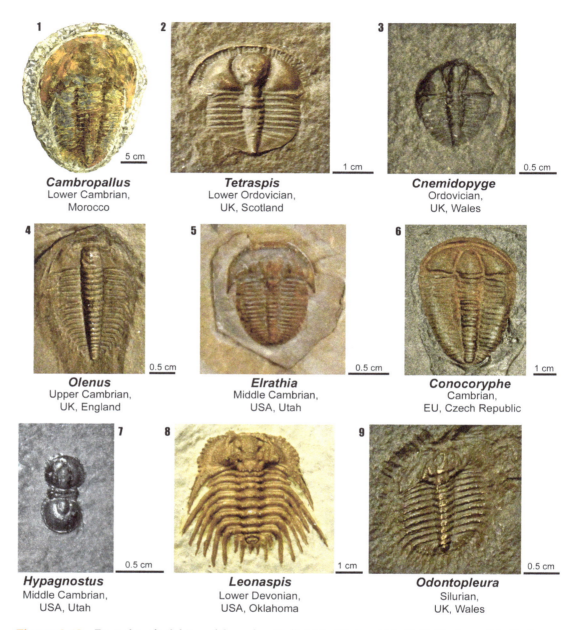

Figure 6.43 Examples of trilobites of the orders Redlichiida (1), Asaphida (2–3), Ptychopariida (4–6), Agnostida (7), and Lichida (8–9). 1: Specimen courtesy of Dr. C. Morgan (University of Calgary); published with permission. 2–9: Specimens from the paleontological collections of the University of Calgary.

6.12 Crustaceans and Chelicerates

Subphylum Crustacea (Cambrian-Holocene) includes primarily marine organisms that colonized brackish and fresh water environments. In many of the taxa of this group one pair of anterior appendages is transformed in claws. Typical crustaceans include lobsters, crabs, shrimps and crayfish. There are different body plans among crustaceans, but in general the cephalon and thorax are fused to form the *cephalothorax*; the posterior component of the carapace of these organisms is the *abdomen*. The organisms of this group are rarely found in the fossil record, and occur

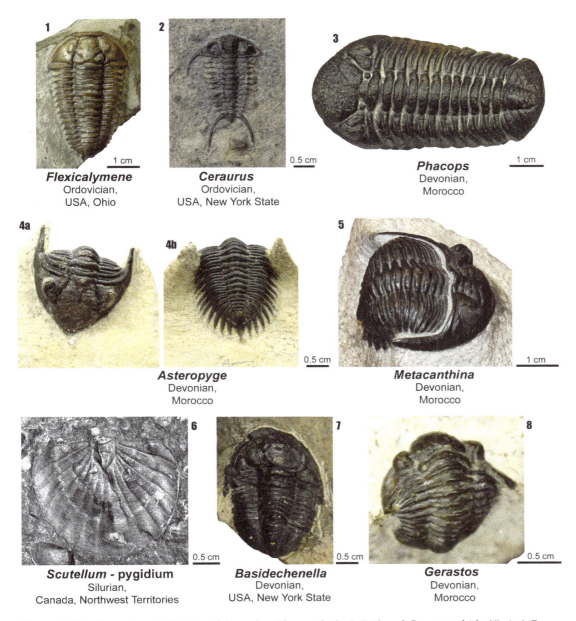

Figure 6.44 Examples of trilobites of the orders Phacopida (1–5, 7–8) and Corynexochida (6). 1, 6–7: Specimens from the paleontological collections of the University of Calgary. 2–5, 8: Specimens courtesy of Dr. C. Morgan (University of Calgary); published with permission.

especially in lagerstätten (Figure 6.45). One particularly interesting group of crustaceans are the cirripeds, which are included within *class Cirripedia*. They are crustaceans that live the life cycle completely attached to the sea floor or floating objects (e.g., whales, ships, etc.). Cirripeds protect their soft body with mineralized plates that can fossilize.

Class Ostracoda includes crustaceans that have the soft body protected by a carapace consisting of two valves, which in the vast majority of taxa are heavily mineralized. Carapace can be easily fossilized and ostracods are among the groups of invertebrates with the highest quality fossil record. Ostracods were primarily marine organisms, but subsequently colonized brackish and fresh water environments; a

Figure 6.45 Examples of fossil crustaceans. Both specimens from the Senckenberg Natural History Museum, Frankfurt; published with permission.

Coeloma
Oligocene,
EU, Germany

Cycleryon
Upper Jurassic, Solnhofen Limestone,
EU, Germany

small number of species adapted even to terrestrial conditions requiring only a small amount of moisture during their life time. The two valves of the ostracod carapace are kept together by a ligament of organic nature and a hinge consisting of an alternance of teeth and sockets; ostracod valve articulation presents significant resemblances with that of the bivalve molluscs. The external surface of the carapace is frequently ornamented with ridges, spines, reticulate structures, etc. (Figure 6.46),

Figure 6.46 Examples of crustaceans of the class Ostracoda.

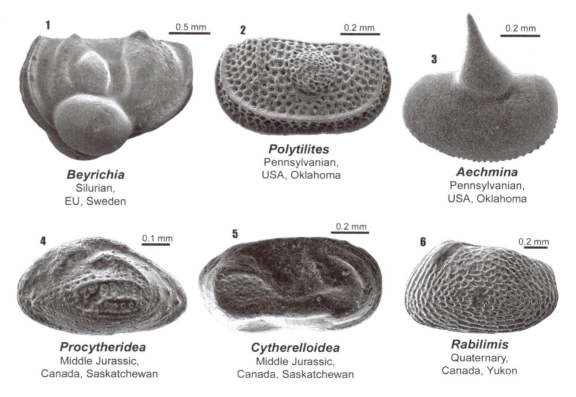

Beyrichia
Silurian,
EU, Sweden

Polytilites
Pennsylvanian,
USA, Oklahoma

Aechmina
Pennsylvanian,
USA, Oklahoma

Procytheridea
Middle Jurassic,
Canada, Saskatchewan

Cytherelloidea
Middle Jurassic,
Canada, Saskatchewan

Rabilimis
Quaternary,
Canada, Yukon

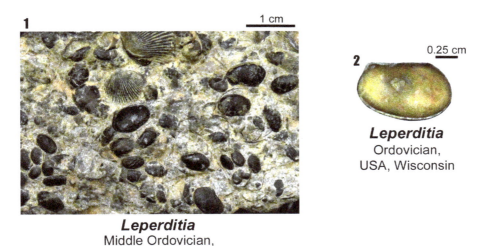

Figure 6.47 Examples of large sized ostracods of the order Leperditicopida. Both specimens from the paleontological collections of the University of Calgary.

features that are extensively used in taxonomy and classification. Stratigraphical range of the ostracods is Cambrian-Holocene and the most apparent trend in their evolution is the calcification of the carapace. In general, the representatives of this class are small-sized taxa, with a carapace length >1 mm. Larger taxa are encountered both in the fossil record and living ostracod assemblages; in one case of gigantism the carapace is as long as 80 mm. Another spectacular example is that of the representatives of the Paleozoic (Ordovician-Devonian) *order Leperditicopida*, which present frequently large-sized carapaces (Figure 6.47).

Subphylum Chelicerata (Ordovician-Holocene) is a diverse group of arthropods adapted to aquatic (marine, brackish, and fresh waters) and terrestrial environments. This subphylum includes organisms such as spiders, scorpions, and horseshoe crabs, which are known both as fossils and living taxa. The representatives of the *subclass Eurypterida* (Ordovician-Permian) were adapted to all the aquatic environments and could move inland over short distances; they are also known under the informal name of *sea scorpions* and some of the gigantic taxa of eurypterids were major predators in the history of life on Earth. A typical eurypterid body plan consists of two segments: the anterior *prosoma* and *opisthosoma* that is situated posteriorly; one of the anterior pairs of appendages is well-developed and was used for pray catching in case they terminated with strong claws or swimming if they evolved in paddle-like structures (Figure 6.48). *Subclass Xiphosura* includes the horseshoe-crabs, and evolved in the Silurian; the highest diversity of the group is known in the Paleozoic, but they declined after the Permian/Triassic boundary. Modern representatives of the horseshoe-crabs are considered living fossils (Figure 2.15).

6.13 Echinoderms

Phylum Echinodermata (Cambrian-Holocene) presents the highest diversity of body plans among the invertebrates, which is reflected in the large number of classes in which it is subdivided. In the recent classification frameworks proposed at the scale of the whole phylum classes are grouped into subphyla based on the occurrence of

Figure 6.48 Examples of chelicerates of the subclass Eurypterida. Both specimens courtesy of Dr. G. Dolphin (University of Calgary); published with permission.

one or a small number of certain morphological features; the resulted subphyla can be regarded at best informal taxonomic units and therefore, this practice is not followed herein.

An overall look at the stratigraphical distribution of the main classes of the phylum Echinodermata indicates some interesting features (Figure 6.49). There is a major diversification during the Cambrian times, when five classes evolved: Eocrinoidea, Helicoplacozoa, Edrioasteroidea, Stylophora, and Homoiostelea. An even higher level of body plan diversification happened during the Ordovician, when seven classes evolved: Cystoidea, Blastoidea, Crinoidea, Ophiuroidea, Asterozoa, Holothuroidea, and Echinoidea. No other echinoderm class evolved after the Ordovician/Silurian boundary. All the echinoderm classes that have representatives in the modern seas and oceans evolved during the second diversification pulse in the Ordovician: Crinoidea, Ophiuroidea, Asterozoa, Holothuroidea, and Echinoidea. There is a certain similarity between echinoderm diversification and that of the trilobites: both groups experienced two successive diversification pulses in the Cambrian and Ordovician; however, trilobites became extinct at the Permian/Triassic boundary, whereas only one echinoderm class became extinct at that time: Blastoidea.

Echinoderm ancestors are not known, but probably should be found among the Ediacaran fossils with elaborate symmetry such as the trilaterally symmetrical *Tribrachidium*. The five classes that evolve during the Cambrian diversification present a wide variability with respect to symmetry.

- *Class Eocrinoidea* (Cambrian-Silurian) consists of taxa that present in general resemblances with the crinoids. The body or *theca* is attached to the sea floor with a *stem* with a root-like structure. Theca consists of plates and pores occur along the suture lines between two adjacent plates. On the theca, and opposed to the attachment point of the stem are the *brachioles*, elongate structures used for catching the food.

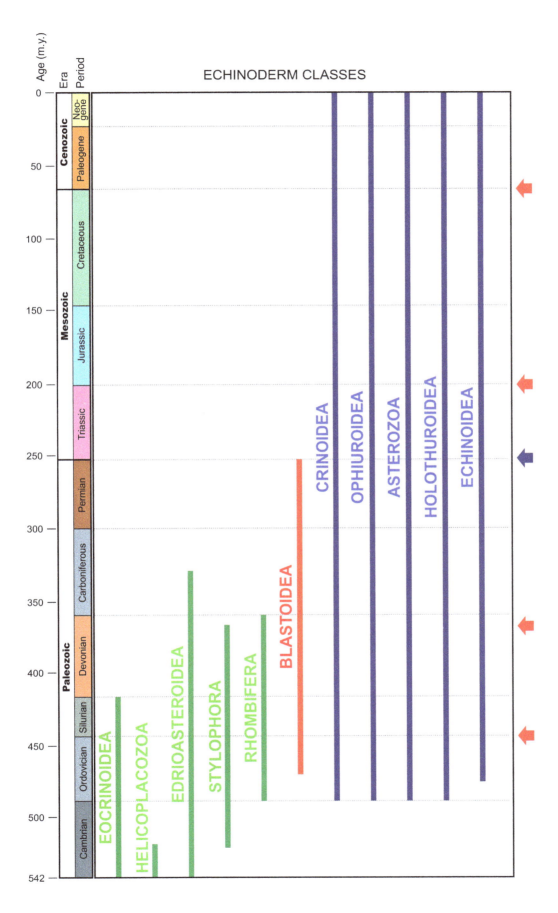

Figure 6.49
Stratigraphical distribution of the classes of the phylum Echinodermata. Green: Paleozoic classes that became extinct before the Permian/Triassic boundary. Red: Paleozoic class that became extinct at the Permian/Triassic boundary; blue-classes that passed the Permian/Triassic major crisis in the history of life and exist in the modern seas and oceans.

The theca has pentameral symmetry or can be flattened, bilaterally symmetrical. *Gogia* is a typical example of eocrinoid (Figure 6.50).

- *Class Helicoplacozoa* (Early Cambrian) presents an ovoid theca with one spirally coiled ambulacrum that can be branched and 10 interambulacral zones. *Helicoplacus* is the only genus included in this class (Figure 6.50).
- *Class Edrioasteroidea* (Cambrian-Early Carboniferous) consists of taxa with pentameral symmetry and a large number of plates forming the theca. The theca is often flattened and the general aspect of the edrioasteroids resembles that of the echinoids indicating a possible ancestor-descendant relationship between the two groups. Edrioasteroids reached the peak diversity in the Ordovician times.
- *Class Stylophora* (Middle Cambrian-Middle Devonian) includes taxa with a flattened larger theca and one anterior arm termed *aulacophore*. Both theca and aulacophore are covered by a large number of plates. Seemingly the stem-like aulacophore was used for movement on the sea floor; if this is a correct interpretation then it can be inferred that stylophorans represent the echinoderm group that achieved probably the highest movement capabilities. *Peltocystis* is an example of stylophoran genus (Figure 6.50).
- *Class Homoiostelea* (Upper Cambrian-Lower Devonian) includes asymmetrical echinoderms that present distinct trends to develop biserial symmetry. The theca is the largest part of the body and is covered by a numerous calcitic plates that do not present an ordered arrangement. There are two extensions from the theca: one posterior (stele) and one anterior (arm). The representatives of this class occur rarely in the fossil record. *Dendrocystites* is a representative genus (Figure 6.50).

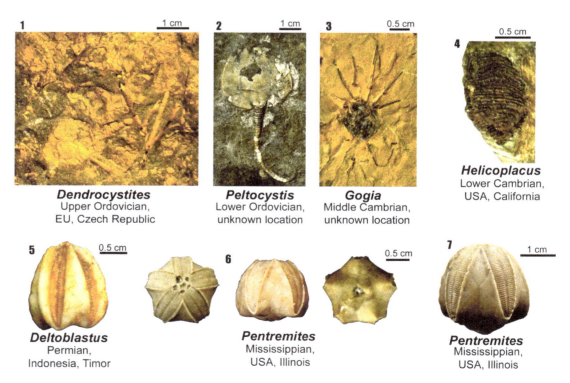

Figure 6.50 Examples of echinoderms of the classes Homoiostelea, Stylopthora, Eocrinoidea, Helicoplacoidea, and Blastoidea. All specimens from the paleontological collections of the University of Calgary.

Two classes that are exclusively Paleozoic evolved as part of the Ordovician echinoderm diversification pulse: Cystoidea and Blastoidea.

- *Class Cystoidea* (Ordovician-Devonian) presents an ovoid theca and the general body architecture resembles that of the crinoids, with thecal pores that in order Rhombifera are arranged in a rhomboidal pattern. The stem occurs in most of the taxa.
- *Class Blastoidea* (Middle Ordovician-Permian) also consists of taxa that present a general crinoid appearance with well-developed pentameral symmetry. The main body or theca (*calyx*) is situated at the end of a stem (*column*), which is attached with root-like structures to the sea floor; brachioles are projected upward from the theca. Theca presents well-developed ambulacral and interambulacral zones. The characteristic respiratory structures of the members of this class are termed *hydrospires*. *Deltoblastus* and *Pentremites* are examples of genera of this class (Figure 6.50).

Of the five classes that evolved in the Ordovician and exist in the modern seas and oceans, the crinoids (*class Crinoidea*) are also important because they are the only echinoderms that contributed to the formation of reefs and this happened in the Late Paleozoic (Figure 6.51). Crinoids reached maximum diversity in the Early Carboniferous and were drastically reduced at the Permian/Triassic major crisis in the history of life. They recovered afterwards slowly with a diversity peak in the Jurassic. Crinoid Holocene diversity is the highest during group's whole evolution.

Figure 6.51 Examples of crinoids of the class Crinoidea. 1–3: Specimens from the paleontological collections of the University of Calgary. 4: Rock slab from the Senckenberg Natural History Museum, Frankfurt; published with permission.

Figure 6.52 Modern representatives of the echinoderm class Asterozoa presenting different types of symmetry. All specimens from the paleontological collections of the University of Calgary.

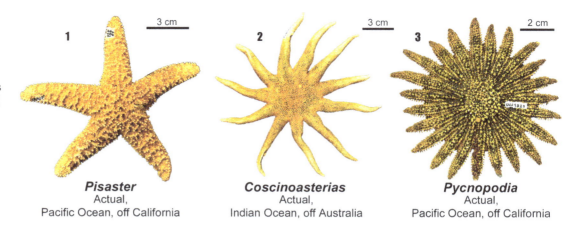

Class Asterozoa were minor components in the echinoderm assemblages until Holocene, when the diversity increased exponentially. One interesting pattern in the asteroid general body plan is the evolution of high-order symmetry in some of the Holocene taxa (Figure 6.52). Notably, a similar stratigraphical distribution pattern of diversity is known in the representatives of the *class Ophiuroidea*.

Class Echinoidea is relatively rare in the fossil record in the earliest portion of the stratigraphic range (Ordovician-Triassic); an accentuated increase in diversity began in the Jurassic and continued till the Holocene; the maximum known diversity is from the Paleogene and Holocene. The most important evolutionary trend recorded in this class is the evolution from the simple pentameral symmetry of the earlier echinoids that are termed *regular* (Figure 6.53) to those in which the symmetry is mixed, namely

Figure 6.53 Examples of regular representatives of the echinoderm class Echinoidea. All specimens from the paleontological collections of the University of Calgary.

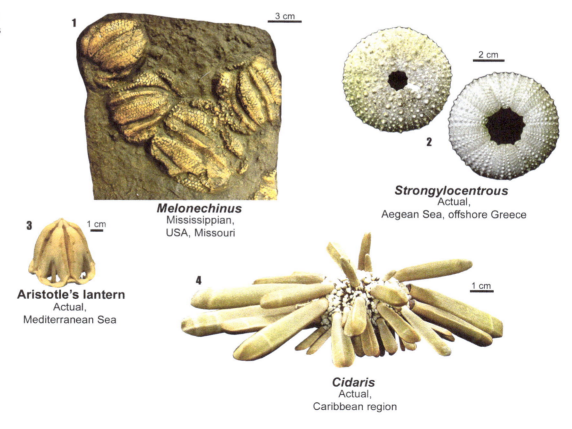

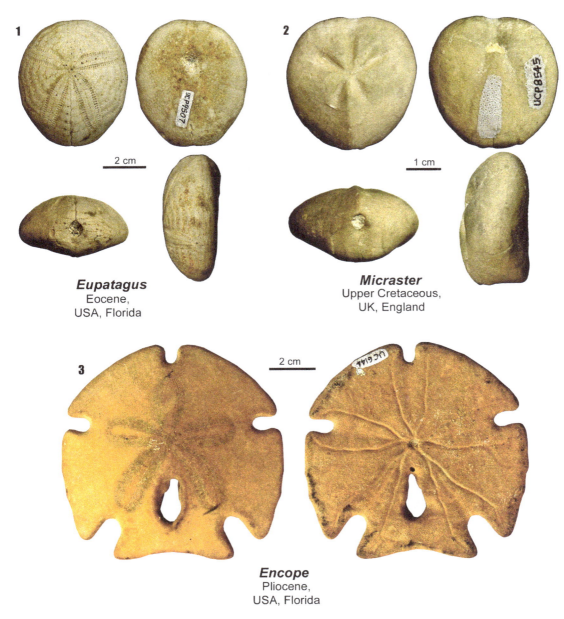

Figure 6.54 Examples of irregular representatives of the echinoderm class Echinoidea. All specimens from the paleontological collections of the University of Calgary.

pentameral with a superimposed bilateral one; taxa of the latter group are also known as *irregular echinoids* (Figure 6.54).

6.14 Graptolites

Graptolites are marine colonial organisms, which are included within class Graptolithina of the invertebrate phylum Hemichordata. Phylum name might be deceiving in indicating an evolutionary relationship with the chordate group; an evolutionary relationship between hemichordates and chordates was never demonstrated. Two other classes are included within the hemichordates: Pterobranchia (e.g., *Cephalodiscus*) and Enteropneusta (accord worms such as *Balanoglossus*). On the overall hemichordates present mixed morphological features between lophophorates (especially bryozoans), echinoderms, and chordates.

Class Graptolithina (Late Cambrian-Early Carboniferous) includes both benthic and planktic taxa; benthic graptolites lived mostly attached to the sea floor. The graptolite colony is termed *rhabdosome* and consists of a variable number of branches, which are known as *stipes*; individual specimens of the colony are situated within small tubular structures opened one end and attached to the stipe on the opposite one; one such tubular structure that hosts one individual is referred to as *theca*. Two orders are recognized within this class: Dendroidea and Graptoloidea.

Order Dendroidea (Late Cambrian-Early Carboniferous) is the older of the two orders and includes taxa with the rhabdosome consisting of a large number of stipes that branch in the direction of growth (Figure 6.55). Originally the dendroid graptolites had a benthic mode of life and by the terminal part of the Cambrian some of the taxa evolved a planktic one. *Rhabdinopora* and *Ptiograptus* are two typical genera of the order Dendroidea. One major evolutionary trend in the graptolite evolution began in the Early Ordovician planktic taxa, and this is marked by a considerable reduction of the number of stipes of the rhabdosome. Such taxa form the *anisograptid* group, which are morphologically intermediary between dendroids and more advanced graptoloids; one genus of anisograptid graptolites is *Aletograptus* (Figure 6.55).

Order Graptoloidea (Ordovician-Early Devonian) includes taxa with the rhabdosome consisting of a small number of stipes and there is a distinct evolutionary trend of stipe number reduction. Graptoloids diversified rapidly in the Ordovician; genera such as *Loganograptus*, *Tetragraptus*, *Plyllograptus*, *Isograptus*, and *Didymograptus* document the earliest diversification of the graptoloids graptolites (Figure 6.56). A new graptoloid group evolved in the Early Silurian; they are the *monograptids* in which the rhabdosome is reduced to one straight or curved stipe only and on it there is only one row of thecae. *Monograptus* and *Cyrtograptus* are typical genera of the monograptid group (Figure 6.57).

Figure 6.55 Examples of graptolites of the order Dendroidea. All specimens from the paleontological collections of the University of Calgary.

6.14 Graptolites

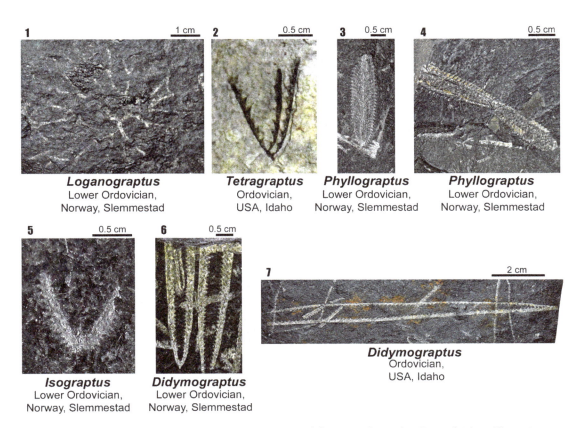

Figure 6.56 Examples of Ordovician representatives of the graptolite order Graptoloidea. All specimens from the paleontological collections of the University of Calgary.

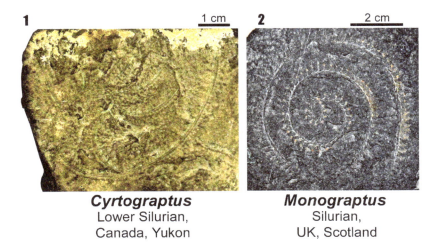

Figure 6.57 Examples of monograptid graptolites. All specimens from the paleontological collections of the University of Calgary.

CHAPTER CONCLUSIONS

- Poriferans are the earliest invertebrates in the fossil record; the evolutionary occurrence of poriferans is in the Cryogenian.
- Evolution of poriferans corresponds to the earliest glaciation pulse of the Snowball Earth; oldest poriferan genus is *Otavia*.
- Archaeocyathid sponges were the earliest reef-building invertebrates in the history of life on Earth.
- Cnidarians evolved in the Ediacaran times; *Cloudina* is the oldest known cnidarian order.
- Orders Tabulata and Rugosa dominate the cnidarian group in the Paleozoic.
- Order Scleractinia evolved after the Permian/Triassic boundary from unknown ancestor and are the dominant organisms in the most complex reef structures in the Earth history.
- Representatives of the class Stenolemata dominated bryozoan assemblages throughout the Ordovician-Cretaceous stratigraphical interval.
- Class Gymnolemata is the dominant living bryozoan group.
- Bryozoans of the class Phylactolemata do not have calcified colonies and occur in fresh and brackish water environments.
- Earliest brachiopods in the fossil record belong to the class Inarticulata.
- Order Orthida is the earliest formal group of articulate brachiopods.
- Living brachiopod assemblages consist of the representatives of orders Rhynchonellida and Terebratulida.
- Modern monoplacophoran genus *Neopilina* documents the body segmentation in molluscs.
- Most of the gastropod taxa are included within subclass Prosobranchia.
- Pteropod gastropods that evolved a planktic mode of life belong to subclass Opistobranchia.
- Gastropods adapted to the terrestrial environments during the Carboniferous times.
- Bivalves are molluscs with bilateral symmetry and with reduced movement capabilities.
- Hippuritoid bivalves of the subclass Heterodonta contributed significantly to the reef formation during the Late Jurassic-Cretaceous times.
- Endoceratoid cephalopods were major predators during the Ordovician.
- Subclass Bactritoidea is the ancestor group for the ammonoids and coleoid cephalopods.
- Lytoceratid ammonoids evolved numerous uncoiled taxa during the Middle Jurassic-Cretaceous times; some of them are excellent index fossils especially for the Cretaceous.
- Rhyncholites and conchorhynchs are calcified jaws of ammonoids and possibly nautilids.
- Mesozoic coileoids present a well-developed rostrum; this structure is considerably reduced or absent in the post Cretaceous representatives of the subclass Coleoidea.

- Trilobites are the oldest arthropods in the fossil record; this group became extinct at the Permian/Triassic boundary.
- Class Ostracoda of the subphylum Crustacea is the arthropod group with the fossil record of highest quality.
- Phylum Echinodermata underwent two major diversification pulses in the Cambrian and Ordovician.
- All the five echinoderm classes that occur in the modern seas and oceans evolved in the Ordovician.
- Class Blastoidea is the only echinoderm class that became extinct at the Permian/Triassic boundary.
- Graptolites were colonial planktic organisms of Late Cambrian-Early Carboniferous age.
- The most apparent trend in graptolite evolution is the reduction of the number of stipes of the rhabdosome.
- Earliest graptolites were benthic organisms; a planktic habitat evolved shortly before the Cambrian/Ordovician boundary.

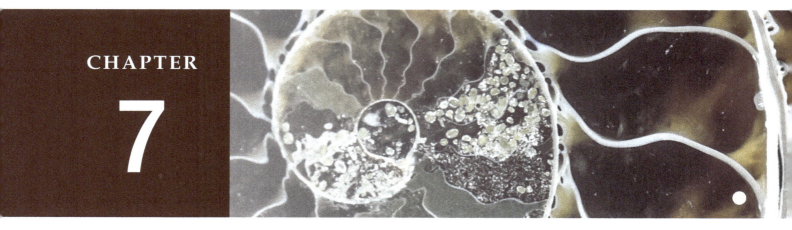

CHAPTER 7

CHORDATE AND VERTEBRATE EVOLUTION

CONTENT

7.1 Chordates
7.2 Agnathan Vertebrates
7.3 Acanthodii, the First Gnathostomes
7.4 Placodermi
7.5 Chondrichthyes
7.6 Osteichthyes
7.7 Amphibians
7.8 Early Reptiles (Carboniferous-Triassic)
7.9 Mesozoic Aquatic Reptiles
7.10 Mesozoic Flying Reptiles
7.11 Dinosaurs
7.12 Birds
7.13 Mammals

Chapter Conclusions

7.1 Chordates

There are two types of animals with axial skeletons: chordates and vertebrates. Chordates have as axial skeleton a rod-like structure termed *notochord*, which is situated in dorsal position is of cartilaginous nature. Vertebrates have the axial skeleton represented by a *vertebral column* that is cartilaginous or of bony nature, consisting mostly of calcium phosphate (apatite), to which various amounts of organic matter are added. Chordates form a group consisting of three phyla: Cephalochordata, Conodonta, and Tunicata, whereas vertebrates are included within phylum Vertebrata.

Phylum Cephalochordata (Cambrian-Holocene) occurs rarely in the fossil record and its representatives are especially found in cases of exceptional preservation. The first chordate reported as fossil was *Pikaia* from the Middle Cambrian Burgess Shale of British Columbia (Canada); *Pikaia* has a small-sized, elongate body with a length of about 5 cm. This organism is morphologically similar in many respects with the modern caphalochordates of the genus *Branchiostoma*; for example there are tentacles around the mouth used for feeding, position of the gill slits just behind the mouth is almost identical in the two genera, just like the V-shaped muscles on the lateral sides, etc. *Pikaia* does not have a well-defined head with sensorial organs such as eyes, etc., and the anterior-posterior axis of the body can be inferred from the position of the mouth and tail. For several decades *Pikaia* remained the only known cephalochordate genus, but this changed with the discovery of the Early Cambrian lagerstätten from Chengjiang (China). By contrast with Burgess Shale, the Chengjiang fauna is considerably richer and more diverse and several genera of cephalochordates and with cephalochordate affinities were described (Figure 7.1). For example, *Yunannozoon* is a genus that resembles cephalochordates and Dzik (1995) suggested that it belongs to a distinct class of chordates: Yunnanozoa. Another genus is *Myllokunmingia* that presents a head with two small eyes situated on the lateral sides of the body; this genus has an anterior skull consisting of cartilaginous tissue. In addition, the lateral body muscles present the same V-shaped pattern also known from *Pikaia* and *Branchiostoma* (Figure 7.1). The morphology of *Myllokunmingia* presents significant differences from the typical cephalochordates and can be considered closer or even a

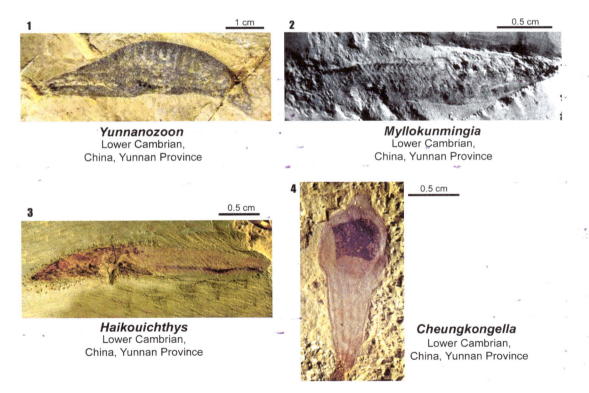

Figure 7.1 Examples of chordates (1–3) and early vertebrates (4). 1: Specimen illustrated by Dzik (1995, fig. 1: A); © Acta Palaeontologica Polonica; published with permission. 2: Specimen illustrated by Hou and Bergström (2003, fig. 20); © Palaeontological Society of Japan; published with permission. 3: Specimen courtesy of Dr. D.-G. Shu (Northwest University, Xi'an, Shaanxi); published with permission. 4: Specimen courtesy of Dr. D.-G. Shu (Northwest University, Xi'an, Shaanxi); published with permission.

full member of the agnathan group and notably, some specialists in the field consider that the taxa of the myllokunmingid group are fishes.

Phylum Conodonta (Late Cambrian-Triassic) evolved from the cephalochordates. Conodont animal body is elongate and in general small-sized, with a length of around 5 cm. Cases of exceptional preservation demonstrated that the axial skeleton of conodont animals consists of a notochord. Two large-sized eyes occur in the head region. Probably the most interesting structure evolved by the representatives of this phylum is the chewing apparatus consisting of small pieces of calcium phosphate and small amounts of organic matter. Conodonts have a fast rate of evolution through the most part of the group stratigraphical range and this is one of the reasons why the pieces of the chewing apparatus are extensively used in biostratigraphy. Conodonts were strongly affected and reduced at the Permian/Triassic boundary, recovered during the Triassic times and became extinct at the Triassic/Jurassic mass extinction event.

Phylum Tunicata (Cambrian-Holocene) consists mostly of solitary taxa, but some colonial species are also known; most of them live attached to the sea floor and a small number of taxa evolved a planktic mode of life. Tunicates have a sparse fossil record, where they occur in cases of exceptional preservation (Shu et al., 2001; Chen et al., 2003) (Figure 7.1) or mineralized spicules that occur in different tissues and organs of the body. One interesting idea is that tunicates evolved in the Ediacaran times (Fedonkin et al., 2012); if this will be supported by further data, then tunicates should be considered the oldest group of chordates.

7.2 Agnathan Vertebrates

Evolutionary occurrence of the vertebrates corresponds with the evolution of the head and vertebral column, which was demonstrated in the case of the Early Cambrian genus *Haikouichthys* of the Chengjiang fauna (Shu and others, 2003) (Figure 7.1). No vertebrates are recorded in the Middle-Upper Cambrian stratigraphical interval. Vertebrates occur again in the fossil record in the Ordovician. All these taxa have in common one important morphological feature: they are jawless organisms and for this reason are included within the *subphylum Agnatha*; living representatives of this subphylum are the hagfishes and lampreys. Hagfishes are marine organisms with eel aspect, well-developed head and rudimentary vertebrae; lampreys are parasitic, a mode of life which was secondarily acquired in the course of evolution. However, fossil agnathans are particularly frequent in the Ordovician-Devonian stratigraphical interval and differ considerably from the modern agnathans. Morphologically, agnathans are extremely diverse and four groups are presented herein: ostracoderms, heterostracans, thelodonts, and osteostracans.

Ostracoderms (Ordovician-Silurian) present the anterior portion of the body covered by bony plates that can consist of a small number of plates, with well-developed dorsal and ventral plates in the case of *Sacabambaspis* or several hundred smaller plates as in *Astraspis*. Ostracoderms had an internal cartilaginous skeleton. The fins are weak or absent, which indicate that ostracoderms were slow swimmers and had a low level of control over the body movements.

Thelodonts (Late Ordovician-Devonian) form a morphologically heterogeneous group of agnathans that have the body covered with characteristic smaller scales, which occur frequently as isolated fossils and in some cases present importance in biostratigraphy. *Furcacauda* and *Loganella* are examples of genera of this agnathan group (Figure 7.2).

Heterostracans (Silurian-Devonian) are extremely diverse from a morphological perspective. They had the anterior portion of the body covered with a multitude of

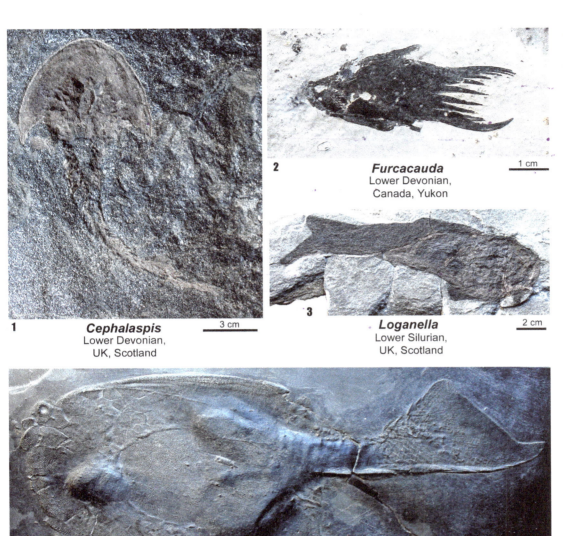

Figure 7.2 Examples of fossil agnathans. All specimens from the collections of the Museum of Natural History, Berlin; published with permission.

bony plates. Some of them, such as *Drepanaspis* lived in the proximity of the sea floor and had the body dorso-ventrally compressed (Figure 7.2). Others evolved a more hydrodynamic form and were faster swimmers.

Osteostracans (Late Silurian-Devonian) evolved a strong bony shield that covered the anterior portion of the body. They evolved higher swimming capabilities as demonstrated by the laterally compressed posterior half of the body and paired fins conferred a better control on the body orientation. Osteostracans were swimmers especially in the proximity of the sea floor. *Cephalaspis* is a classic example of genus of this group (Figure 7.2); two sensorial zones developed on the cephalic shield were probably used to recognize water vibrations.

Ordovician-Devonian is the stratigraphical interval in which agnathans are frequently encountered in the fossil record. After the Devonian/Carboniferous boundary these organisms entered in a long period of decline of over 360 million years, which continues today.

7.3 Acanthodii, the First Gnathostomes

A major leap in the evolution of vertebrates is represented by the development of jaws. This event is documented in the Early Silurian times and marks the evolutionary occurrence of the *subphylum Gnathostomata*. The term Gnathostomata is translated as "jawed mouth" and includes superclass Pisces, class Amphibia, class Reptilia, class Aves, and class Mammalia. The earliest group of gnathostomes are the acanthodians, which are formalized as class within superclass Pisces.

Class Acanthodii (Silurian-Early Permian) have certain resemblances in the general body aspect with the sharks, present a cartilaginous skeleton and this indicates their primitive nature; the representatives of this class were sometimes referred to as "spiny sharks." There are very few bony structures in an acanthodian skeleton. They are particularly important and the name of the whole group is derived from them. The basal structures of each of the paired fins (pectorals and abdominal) as well as the dorsal one are bony; in addition, each of these fins presents anteriorly a bony spine that can be easily recognized in fossil material (Figure 7.3). Acanthodians have the body covered with small-sized scales that resemble morphologically those that occur in some bony fishes and did not

Figure 7.3 Examples of acanthodians gnathostomes. All specimens from the collections of the Museum of Natural History, Berlin; published with permission.

Acanthodes
Lower Permian,
EU, Germany

Diplacanthus
Middle Devonian,
UK, Scotland

Gyracanthus
Upper Carboniferous,
UK, England

evolve a cephalic shield. The representatives of this class were marine at the beginning of their evolution but in the Devonian times a considerable number of species adapted to fresh water environments; this event can be correlated with the evolution of the Devonian elasmobranch and placoderms predators. Larger acanthodians are known from the later part of the stratigraphical range of the group; such organisms were up to 3 m in length and are often documented by the fossilized spines. Examples of acanthodian genera are *Acanthodes*, *Diplacanthus*, and *Gyracanthus* (Figure 7.3).

7.4 Placodermi

Class *Placodermi* (Latest Silurian-Devonian) is the first class of fishes that evolved major predators in the history of life and they were the dominant fish group during Devonian times; the group evolved in the Late Silurian and the diversification pulse during the Earliest Devonian led to the occurrence of most of the placoderm subgroups. Placoderms have a cartilaginous and partly ossified axial skeleton and the anterior portion of the body covered by articulated bony plates, hence their name; based on this general aspect the representatives of class Placodermi present certain resemblances with the older ostracoderms and heterostracans; the occurrence of jaws in placoderm taxa differentiates them from the two agnathan subgroups. They were fishes adapted to marine and fresh water environments and occasionally evolved large sizes. Two large groups of placoderms that include most of the species of this class are recognized and formalized at order level: Antiarchi and Arthrodira.

Order *Antiarchi* (Late Silurian-Devonian) are in general small-sized and with a well-developed cephalic shield consisting of large ornamented bony plates; in addition, bony plates covered the pectoral fins, which in many taxa are elongate. Antiarchs are taxa that dominated in fresh water environments. In general it is accepted that antiarchs were slow-swimmers. Examples of genera of the order Antiarchi are *Coccosteus*, *Rachiosteus*, *Bothriolepis*, and *Pterichthyodes* (Figure 7.4).

Order *Arthrodira* (Devonian) were mostly marine but some taxa adapted to fresh water environments. They were major predators that reached large sizes; for example, *Dunkleosteus* of the Late Devonian could be as long as 6 m and had a massive cephalic shield with the component plates articulated in points conferring thus a high level of mobility (Figure 7.5). Large sizes are also encountered in the freshwater representatives of this group; *Tityosteus* of the Early Devonian was over 2 m in length.

One interesting example of placoderm fish is represented by *Gemündina*, which is included in a smaller order and was originally described from the Lower Jurassic rocks of Germany (Figure 7.4). It was a smaller predator with dorso-ventrally flattened body covered small unfused plates; this is a primitive feature that contrasts with the cephalic shield plates of the evolved orders Antiarchi and Arthrodira.

7.5 Chondrichthyes

Class *Chondrichthyes* (Late Silurian-Holocene) includes fishes with cartilaginous skeleton and the only bony parts are represented by the teeth, which occur frequently in the fossil record both as macrofossils and in micropaleontological samples. Representatives of this class are often referred to informally as *cartilaginous fishes*. Notably,

Figure 7.4. Examples of placoderms of the orders Antiarchs (1–3, 5) and Rhenanida (4). All specimens from the collections of the Museum of Natural History, Berlin; published with permission.

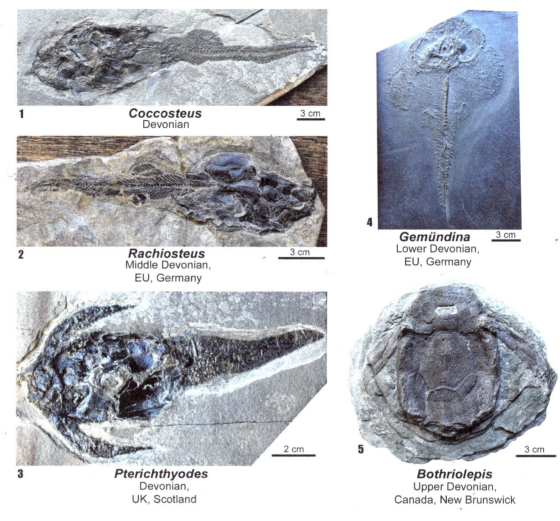

1 *Coccosteus* — Devonian — 3 cm

2 *Rachiosteus* — Middle Devonian, EU, Germany — 3 cm

3 *Pterichthyodes* — Devonian, UK, Scotland — 2 cm

4 *Gemündina* — Lower Devonian, EU, Germany — 3 cm

5 *Bothriolepis* — Upper Devonian, Canada, New Brunswick — 3 cm

Figure 7.5. Example of a cephalic shield of one placoderm of the order Arthrodira. Specimen from the Museum of Natural Sciences, Paris; photographed by the author, August-2014.

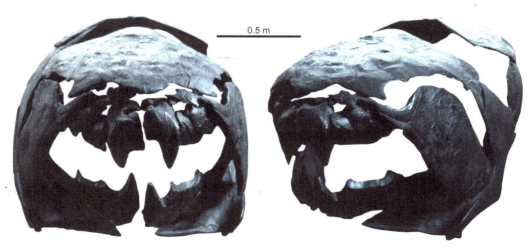

Dunkleosteus - cephalic shield reconstruction
Upper Devonian, USA — 0.5 m

there are species of this class in which certain components of the skeleton, other than the teeth and especially vertebrae can present an incipient ossification. This class includes fishes such as sharks, rays chimaeras. The body is covered by a rough skin consisting of small sclerites; scales are not known in the representatives of this class. The vast majority of the class Chondrichthyes were marine organisms and only in rare cases colonized lower salinity environments.

There are two major body plans among the representatives of the class Chondrichthyes. One is hydrodynamic and is best exemplified in the case of sharks. Taxa with such body architecture are excellent swimmers and many of them ferocious predators in shallow marine waters and deep oceanic ones. This group evolved rapidly and among the most frequent representatives in the Paleozoic times are *Cladoselache* (Late Devonian) and *Stethacanthus* (Late Devonian-Early Carboniferous). *Lebacanthus* of the Carboniferous-Early Permian is an example of one shark genus adapted to the fresh waters (Figure 7.6). The largest shark genus known is *Carcharodon* of the Neogene, which can be as long as 12 m.

The other group of Chondrichthyes consists of taxa with the body strongly compressed dorso-ventrally; they are often referred to as *batoids* or *batoid sharks*. They are most frequently encountered in the proximity of the sea floor and occur in the fossil record mostly as detached teeth; complete skeletons occur only in cases of exceptional preservation. *Squatina* and *Spathobatis* are examples of genera of this group (Figure 7.6).

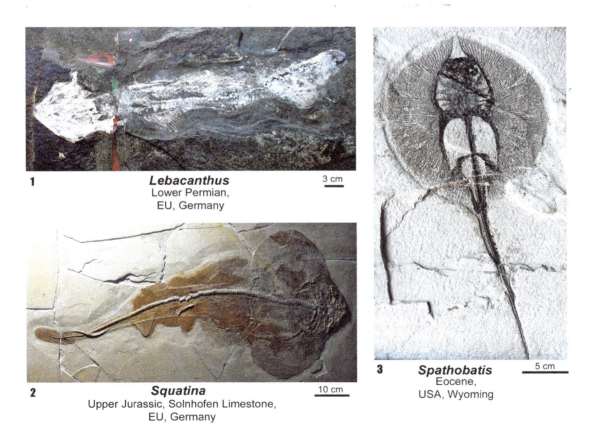

Figure 7.6 Examples of fishes of the class Chondrichthyes. 1, 3: Specimens from the collections of the Museum of Natural History, Berlin; published with permission. 2: Specimen from the Senckenberg Natural History Museum, Frankfurt; published with permission.

7.6 Osteichthyes

Class Osteichthyes (Late Silurian-Holocene) includes fishes with a completely ossified skeleton, which are also known as bony fishes. They are excellent swimmers and control the body position in the water with the aid of two pairs of fins (pectoral and abdominal) and a variable number of unpaired ones; of the unpaired ones the *caudal fin* is the most important for it represents the main organ used in locomotion and maneuvering. The head is covered by wide bony plates that have a protective role, and the rest of the body is covered by scales. Bony fishes earliest occurrence is known from the Late Silurian and diversified rapidly in the Early Devonian times. Class Osteichthyes is subdivided into two subclasses based on the fin morphology: Actinopterygii and Sarcopterygii.

Subclass Actinopterygii (Late Silurian-Holocene) includes bony fishes in which the fins consist of skin tissue with thin bones or rays radiating from the fin base (Figure 7.7); the representatives of this subclass are also known as *ray fin fishes*. The group earliest pulse of increasing diversity is encountered in the Devonian times, shortly after their evolution and in a time in which they lived in hostile environments dominated by the massive placoderm predators. Actinopterygian diversification began in the Early Carboniferous, shortly after the placoderm predator's demise in the proximity of the Devonian/Carboniferous boundary. Since the beginning of their evolution, actinopterygians were fast swimmers and many species developed large populations, a very effective strategy for species survival. Earliest actinopterygians were marine fishes but in time colonized brackish and fresh water environments; group diversity increased gradually and in the Holocene times they are the most diverse group of fishes. One spectacular achievement of the group is the evolution of flight during the Triassic times. Fish flight is realized with the aid of well-developed pectoral fins and is over short distances of several tens of meters; *Exocoetus* is an example of genus of flying fish.

Subclass Sarcopterygii (Late Silurian-Holocene) includes taxa in which the fins have a basal portion with an extension of the body fleshy tissues supported by strong

Thrissops
Upper Jurassic,
Solnhofen Limestone
EU, Germany

Diplomystus
Upper Jurassic,
Solnhofen Limestone
EU, Germany

Figure 7.7 Examples of fossil bony fishes of the class Osteichthyes, subclass Actinopterygii. Both specimens from the Senckenberg Natural History Museum, Frankfurt; published with permission.

bones and a distal part that presents similarities with the actinopterygians and consists of skin tissue supported by bony rays. Due to the fin morphology, sarcopterygians are also known as *lobe fin fishes*. The earliest occurrence of the sarcopterygians is known from the Late Silurian of China (Zhu and others, 2009). These fishes were primarily marine, and most of the taxa remained confined to this environment throughout the group's evolution; other colonized fresh water environments and a smaller number evolved capability to breathe atmospheric oxygen. Sarcopterygii are the fishes that colonized terrestrial environments and evolved into amphibians during the Late Devonian times. Three orders are included in this subclass: Actinistia, Dipnoi, and Rhipidistia.

Order Actinistia (Devonian-Holocene) includes the typical coelacanths (Figure 7.8). A long time it was believed that the representatives of this order became extinct at the Cretaceous/Paleogene boundary, but the living species *Latimeria chalumnae* was discovered in the 1930s in the western Indian Ocean.

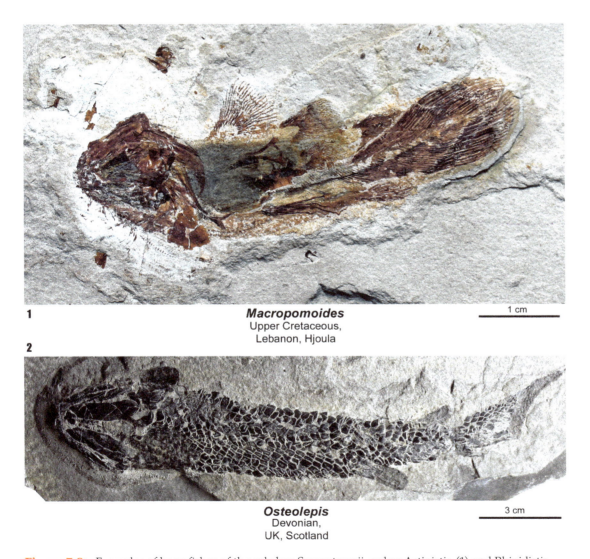

1 ***Macropomoides***
Upper Cretaceous,
Lebanon, Hjoula

2 ***Osteolepis***
Devonian,
UK, Scotland

Figure 7.8 Examples of bony fishes of the subclass Sarcopterygii, orders Actinistia (1) and Rhipidistia (2). Both specimens from the collections of the Museum of Natural History, Berlin; published with permission.

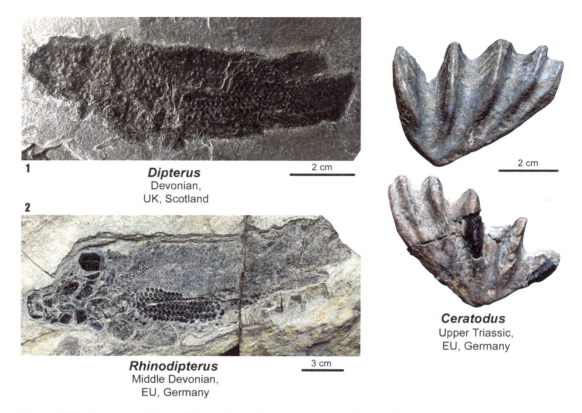

Figure 7.9 Examples of bony fishes of the subclass Sarcopterygii, order Dipnoi. Both specimens from the collections of the Museum of Natural History, Berlin; published with permission.

Order Dipnoi (Devonian-Holocene) consists of taxa adapted to fresh waters, evolved incipient lungs and the capability to breathe atmospheric oxygen. This adaptation is paramount in surviving the frequent periods of drought in tropical regions. Dipnoans are also known as *lungfishes*. *Dipterus* and *Rhinodipterus* are typical Devonian representatives of the group (Figure 7.9). One case of gigantism among dipnoans is recorded in the Triassic-Paleogene genus *Ceratodus*. There are three living genera of dipnoans, all of them in continents of the southern hemisphere and endemic for one particular continent: *Neoceratodus* in Australia, *Lepidosiren* in South America, and *Protopterus* in Africa.

Order Rhipidistia (Devonian-Early Permian) is that group of sarcopterygians from which land tetrapods evolved in the Late Devonian times. They inhabited mostly the shallow marine environments. *Osteolepis* and *Eusthenopteron* are two examples of genera of this order (Figure 7.8).

7.7 Amphibians

Class Amphibia (Late Devonian-Holocene) evolved as the result of land invasion by fishes. The comparison between the axial skeleton and head bone arrangement of the two groups indicates that the earliest land vertebrates evolved from rhipidistian lobe-finned fishes. Recent studies focused on the fish-to-amphibian evolution showed that the lobe-finned genus *Eusthenopteron* is most likely the direct ancestor of the earliest land tetrapods. The earliest amphibians are of Late Devonian age and are assigned to two genera: *Acanthostega* and *Ichthyostega*. Both have well-developed limbs and a strong tail; skeleton morphology indicates that none of them was completely adapted to terrestrial environments and developed much of the life cycles in aquatic conditions. The stronger limbs of *Ichthyostega* when compared to those of *Acanthostega* suggest that it could spend more time in terrestrial conditions. Evolution from fishes to amphibians is reflected by major changes in skeleton morphology. One of the most important is the transformation of the fins into limbs and evolution of toes. *Acanthostega* and *Ichthyostega* had six or seven toes; genus *Pederpes* of the early Carboniferous is the first land vertebrate that evolved a limb with five toes; this remained a fingerprint in the terrestrial vertebrates although modifications from this type are often encountered both in the fossil record and modern faunas.

Respiration strategy and mechanism changed significantly with the adaptation to the terrestrial conditions. Amphibians have well-developed lungs, which provide a part of the necessary oxygen; the other part is taken through skin (*cutaneous respiration*) and for this reason amphibians must avoid a complete drying of the skin. Amphibians lack the chest bone and consequently do not breathe actively by periodically increasing and decreasing the thoracic cavity volume as in the more advanced vertebrate groups (e.g., reptiles, mammals, birds). Reproduction is through eggs, one strategy that has many similarities with that known in the fish ancestors; the large numbers of eggs are unprotected, laid in aquatic environments and are not nurtured. The earliest amphibians have skulls that resemble in many respects those of the ancestral rhipidistian; there are no openings behind the eye sockets and this kind of skull is considered of *anapsid type*. These amphibians are often referred to as *basal tetrapods* and are the ancestors of two other large groups of fossil amphibians formalized at order level: Temnospondyli and Anthracosauria.

Order Temnospondyli (Late Carboniferous-Early Cretaceous) is diverse from a morphological perspective. Some of the members of this group have crocodile-like aspect (e.g., *Sclerocephalus, Capitosaurus, Eryops*, etc.), with large skulls and strong jaws, others resemble smaller salamanders (e.g., *Apateon, Micromelerpeton*, etc.) (Figure 7.10). Group's evolution shows a gradual adaptation to the terrestrial environments. Temnospondyls, which are often referred to as *stegocephalians* and *labyrinhodonts*, are the ancestors of the modern amphibians or lissamphibians that are encountered during the Permian-Holocene stratigraphical interval and include frogs, newts, salamanders, etc. Lissamphibians are relatively rare occurrences in the fossil record (Figure 7.11).

Order Anthracosauria (Carboniferous-Permian) are the amphibians that will eventually evolve into reptiles. Most of the anthracosaurs were adapted to life in aquatic conditions and only a few were completely adapted to terrestrial environments. *Seymouria* of the Permian of Germany is one of the best known genera of anthracosaurs. The group became extinct at the Permian/Triassic boundary.

7.8 Early Reptiles (Carboniferous-Triassic)

Class Reptilia (Carboniferous-Holocene) are the first vertebrate organisms in the history of life on Earth that evolved active breathing and experimented considerably with the mechanisms of thermoregulation. Primarily they were terrestrial organisms but some groups adapted to aquatic environments of all types (marine, brackish and fresh waters) and iteratively evolved flight. Paleoenvironments in which reptiles lived changed in certain groups and for example, the genera *Mesosaurus* and *Stereosternum* of the Permian times, which are found in the continents of the southern

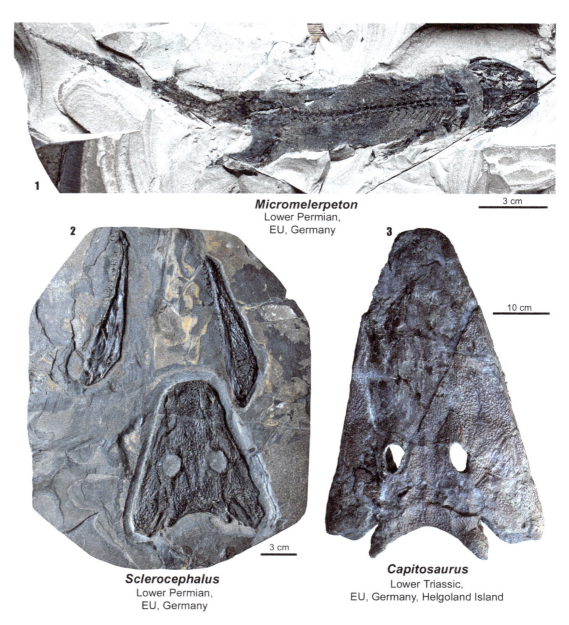

Figure 7.10 Examples of amphibians of the order Temnospondyli. All specimens from the collections of the Museum of Natural History, Berlin; published with permission.

hemisphere, were among the earliest reptiles that readapted to low-salinity aquatic environments (Figure 7.12). Evolution from amphibians to reptiles was a major evolutionary leap and the transition to higher vertebrates is apparent not only in morphological changes (e.g., evolution of chest bone) but also changes in the reproduction mechanism. The higher vertebrates include reptiles, mammals, and birds; these organisms are grouped together as amniotes. *Amniotes* have the embryos protected by eggs in which the soft tissues and substances are protected by a hard mineralized shell; the shell encloses the embryo and nutrients necessary for the young individual. By contrast to fishes and amphibians, amniotes lay a smaller number of eggs, which are nurtured in the evolved species.

The earliest reptiles evolved in Carboniferous. They were small-sized lizard-like organisms with the skull having two openings behind the eye socket (*diapsid type*).

7.8 Early Reptiles (Carboniferous-Triassic) 181

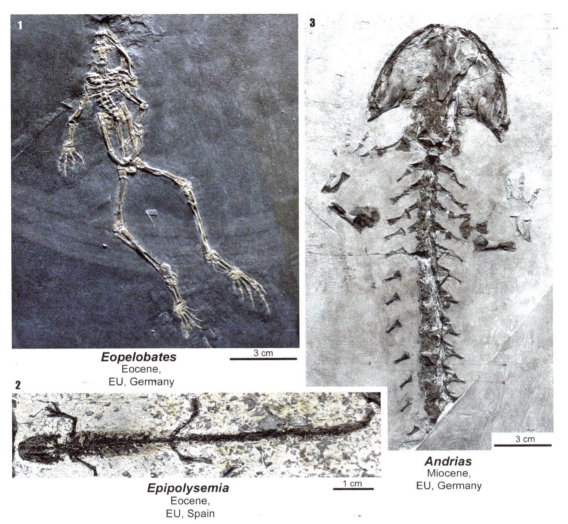

Figure 7.11 Example of fossil lissamphibians. 1–2: Specimens from the Senckenberg Natural History Museum, Frankfurt; published with permission. 3: Specimen from the collections of the Museum of Natural History, Berlin; published with permission.

Figure 7.12 Examples of Late Paleozoic reptiles. Both specimens from the Senckenberg Natural History Museum, Frankfurt; published with permission.

Petrolacosaurus and *Hylonomus* of this period are two examples of primitive (basal) diapsid reptiles. The dominant Late Paleozoic reptiles of the Late Carboniferous-Permian times were of different kind; they had only one opening behind the eye socket (*synapsid type*). The earliest reptile adaptive radiation resulted in the evolution of *pelycosaurs*; this group evolved the first reptilian mechanism of body temperature regulation realized through highly vascularized skin on the dorsal part of the body. Sail backs were used to capture solar energy for body heating. Pelycosaurs were both vegetarians (e.g., *Edaphosaurus*) (Figure 7.13) and carnivores (e.g., *Dimetrodon*). *Dimetrodon* was the top predator during Early Permian terrestrial environments. Pelycosaurs (sailback reptiles) lived in regions with warm climate: equatorial and tropical.

Formation of the supercontinent Pangea allowed the pelycosaurs to colonize temperate and cold climate regions in both north and south hemispheres. A new type of synapsid reptiles evolved in the early Permian, namely the therapsids (*order Therapsida*), which are also known as mammal-like reptiles. Some therapsids species evolved large-sized bodies with a thick layer of fat tissue, an adaptation to retain the body heat in colder climate. It is possible that some of these reptiles to have had warm blood and body covered with fur. Therapsids, which flourished in the Permian when they were the dominant terrestrial reptile group, survived the Permian/Triassic boundary crisis. They are subdivided into several subgroups, some of them carnivorous specialized on different kinds of prey (*gorgonopsians*), others vegetarians that lack teeth (*anomodonts*) and others with mixed feeding regime (*therocephalians*). One group of therapsids known as cynodonts are of particular importance due to their morphologic resemblances with the mammals. They evolved in the supercontinent Gondwana and reached global distribution in the Early Triassic. *Cynodonts* were small-sized animals in which the dentition was further developed and has significant resemblances with those of mammalian type; they are considered the ancestors of the *mammals*, the latter having the evolutionary occurrence in the Late Triassic. The most significant change in the reptilian faunas of the Triassic was the overall replacement of the synapsids by diapsids. The oldest diapsid genus is *Petrolacosaurus*, a small-sized insectivore. Diapsids were not among the flourishing and dominant reptiles of the Carboniferous and Permian, but colonized a variety of environments, and adopted different feeding

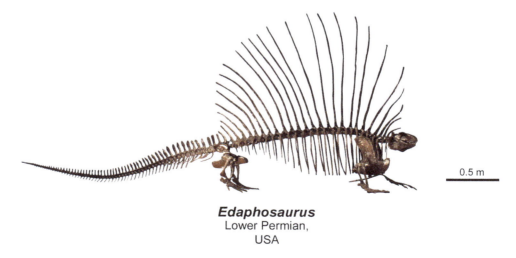

Edaphosaurus
Lower Permian,
USA

Figure 7.13 Example of a pelycosaur. Specimen exhibited in the National Museum of Nature and Science, Tokyo; base photograph courtesy Dr. K. Tanaka; published with permission.

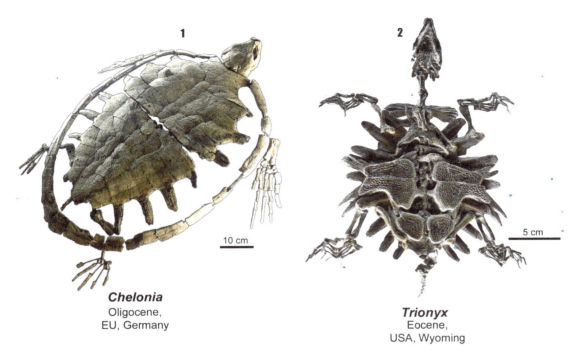

Figure 7.14 Examples of representatives of the order Testudines. Both specimens from the Senckenberg Natural History Museum, Frankfurt; published with permission.

strategies [insectivores (e.g., *Petrolacosaurus*), carnivores (e.g., *Heleosaurus*), etc.] and adapted to various ecological niches [tree gliders (e.g., *Coelurosauravus*), aquatics (e.g., *Hovasaurus*), etc.]. Diapsid diversification began in the Late Permian and two major groups evolved in the Triassic: lepidosaurs (e.g., lizards, snakes, etc.), and archosaurs (e.g., crocodiles, etc.).

Order Testudines (Triassic-Holocene) are anapsid reptiles forming a polyphyletic group. The anapsid condition is achieved through simplification from the diapsid one and seemingly this happened iteratively several times in the course of reptile evolution. Turtles have the body protected by a shell consisting of a *carapace* (upper portion) and a *plastron* (lower part); the carapace and plastron are fused on the lateral sides leaving space for legs and necks (Figure 7.14). The most advanced turtles can retreat completely within the shell in case of danger.

7.9 Mesozoic Aquatic Reptiles

Throughout the Mesozoic times, reptiles adapted several times to the aquatic conditions. Seas and oceans offered a hospitable environment for the reptile adaptation primarily due to the gradually increasing organic productivity; practically all the ecologic niches in the Mesozoic seas and oceans could yield the necessary food required for the evolution of predator species, which at times were large-sized.

Order Ichthyosauria (Triassic-Cretaceous) can be easily recognized by the fish-like or dolphin-like body shape. The head is elongate, and presents two large-sized eyes on each side. The tail is developed in vertical plane as in fishes rather than horizontally as in dolphins. The representatives of the ichthyosaurid group are the reptiles best adapted to the life in marine environments; they were fast swimmers and could inhabit ecological niches at various depths. Ichthyosaurs gave birth to living offspring with birth mechanisms that present considerable resemblances with that of the dolphins. *Leptonectes* of the Early Jurassic times is an example of ichthyosaurid genus (Figure 7.15).

Order Sauropterygia (Triassic-Cretaceous) are marine reptiles that include the *nothosaurs* of the Triassic and *plesiosaurs* of the Jurassic and Cretaceous. These reptiles had an elongate hydrodynamic body and a small head situated at the end of a long and thin neck. Nothosaurs had distinct limbs and these parts of the body evolved into paddles in plesiosaurs, an adaptation that also occurs in the ichthyosaurid group. *Futabasaurus* of the Late Cretaceous is an example of plesiosaurid genus (Figure 7.15).

Order Placodontia (Middle-Late Triassic) form a small aquatic reptilian group whose representatives inhabited shallow water environments from the proximity of the shoreline. Placodont dentition is adapted for crushing the mollusc shells, such as bivalves, gastropods, and possibly cephalopods. A classic example of placodont genus is *Placodus* (Figure 7.16).

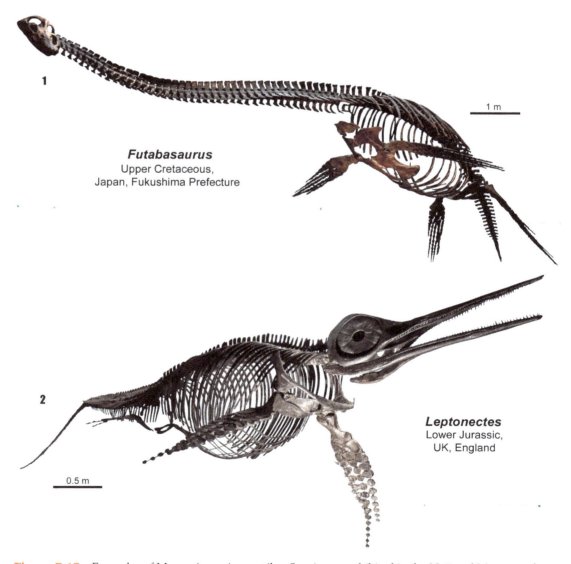

Figure 7.15 Examples of Mesozoic marine reptiles. Specimens exhibited in the National Museum of Nature and Science, Tokyo; base photographs courtesy Dr. K. Tanaka; published with permission.

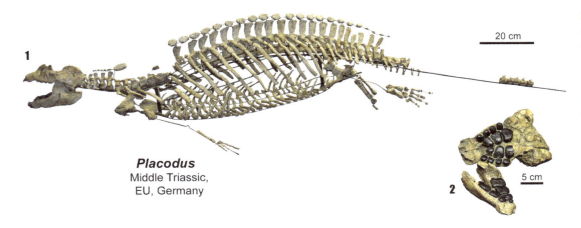

Figure 7.16 Example of representatives of the order Placodontia. Both specimens from the Senckenberg Natural History Museum, Frankfurt; published with permission.

7.10 Mesozoic Flying Reptiles

Reptiles developed flight during iteratively several times during the Mesozoic. No flying reptiles survived the Cretaceous/Paleogene boundary mass extinction. They evolved from archosaurians and have no direct evolutionary relationship with the gliding lizard-like reptiles of the Permian and Early Triassic.

Order Pterosauria (Late Triassic-Cretaceous) includes about 100 species of flying reptiles. Group's name literally means *winged reptiles*. Genus *Rhamphorhynchus* of the Jurassic is a classic example of a pterosaurian reptile (Figure 7.17). The representatives of this order were considered in the past gliders from the higher places; the small-sized species used most likely the trees to start the gliding, whereas the larger ones started their glided flight from a cliff. Pterosaurian flight capabilities were re-evaluated and it was demonstrated that some species could have moved actively their wings during the flight, making them highly maneuverable in the air; these animals were capable of using the air currents to increase the duration and distance of the flight and at least in the Cretaceous they were capable to cross the Atlantic Ocean, which was much narrower at that time. The efficiency of the pterosaurian flight can be compared to that of the birds today. In contrast, these reptiles were inefficient as they moved on the ground.

Figure 7.17 Example of a flying reptile. Specimen exhibited in the Museum of Natural History, Berlin; published with permission.

The adaptations of the pterosaurian reptiles for flight are evident, and the occurrence of such morphologic features leaves no doubt on their way of life. Pterosaurians had hollow bones; these bones are lighter and occur in all the groups that evolved flight capabilities. The heads throughout the pterosaurian group are streamlined; such aerodynamic form was an advantage during flight and especially in the case of the animals that evolved the glided flight. Pterosaurian body was covered with hair as demonstrated for the first time in the case of *Sordes pilosus*, a species described from the late Jurassic of Kazakhstan. Bodies covered with hair are a clear indication that the members of this order or at least some of them evolved *endothermy*; therefore, their blood should have been much warmer than that in the contemporaneous reptile groups. Wings consisted of extended skin, which was attached to the modified arm skeletal elements over the whole length; the wings could be moved through the action of the massive pectoral muscles.

7.11 Dinosaurs

Superorder Dinosauria (Late Triassic-Cretaceous) is a group of Mesozoic reptiles, which evolved in the Triassic from thecodont ancestors; they are typically diapsid. In the Jurassic and Cretaceous the dinosaurs were the dominant land vertebrates and their evolution resulted in the occurrence of the some of the largest animals in the life history on Earth. A fundamental structure in the dinosaurian skeleton is the form and orientation of the pubis bone; this feature, which has a significant contribution in dinosaurian evolution and classification. This feature is used to recognize two groups of dinosaurs, which are formalized at order level: *Saurischia* and *Ornithischia* (Figure 7.18).

Order Saurischia (Late Triassic-Cretaceous) presents the *pubis* bone oriented downward and forward from the articulation with the *ilium* and *ischium*, the other two bones that form the pelvic girdle. This order is subdivided into three suborders: Staurikosauria, Theropoda, and Sauropodomorpha. The name indicates that the pelvic girdle has resemblances with that of the modern lizards.

Order Ornitischia (Late Triassic-Cretaceous) has the pelvic girdle with the pubis bone in ventral position and parallel to the ischium that is extended backwards. There are five suborders included in this order: Ornithopoda, Pachycephalosauria, Stegosauria, Ankylosauria, and Ceratopsida. The name of the order is derived from the resemblances between their pelvic girdle with that of the birds.

It was considered in the past that dinosaurs evolved in the Late Triassic (Carnian). Trace fossils suggest that the dinosaurid group evolved earlier, namely in the latest Middle Triassic (late Ladinian). However, skeletons with predinosaurian features, which are included in the informal *dinosauromorph* group, are known from the early middle Triassic (Anisian); track fossils of the dinosauromorph group are found in sedimentary rocks as old as Early Triassic (Olenekian) (Figure 7.18). Therefore, our data indicate that the representatives of the archosaurian group began the evolution toward the dinosaurs during the Early Triassic. Dinosaurids evolved rapidly in the Late Triassic and the earliest diversification pulse is documented in the Late Triassic (Norian). The earliest dinosaurs were bipedal small carnivores, which ranged in length from one half meter to about 4 m; two such Late Triassic earlier dinosaurs are *Staurikosaurus* from Brazil, and *Herrerasaurus* of Argentina (Figure 7.19). Their pelvic girdle presents resemblances with those of the thecodont ancestors.

Saurischian dinosaurs evolved in the Triassic and all the three suborders have the evolutionary occurrence during this period. Among them, the most primitive ones that are included in *suborder Staurikosauria* did not survive the Triassic/Jurassic

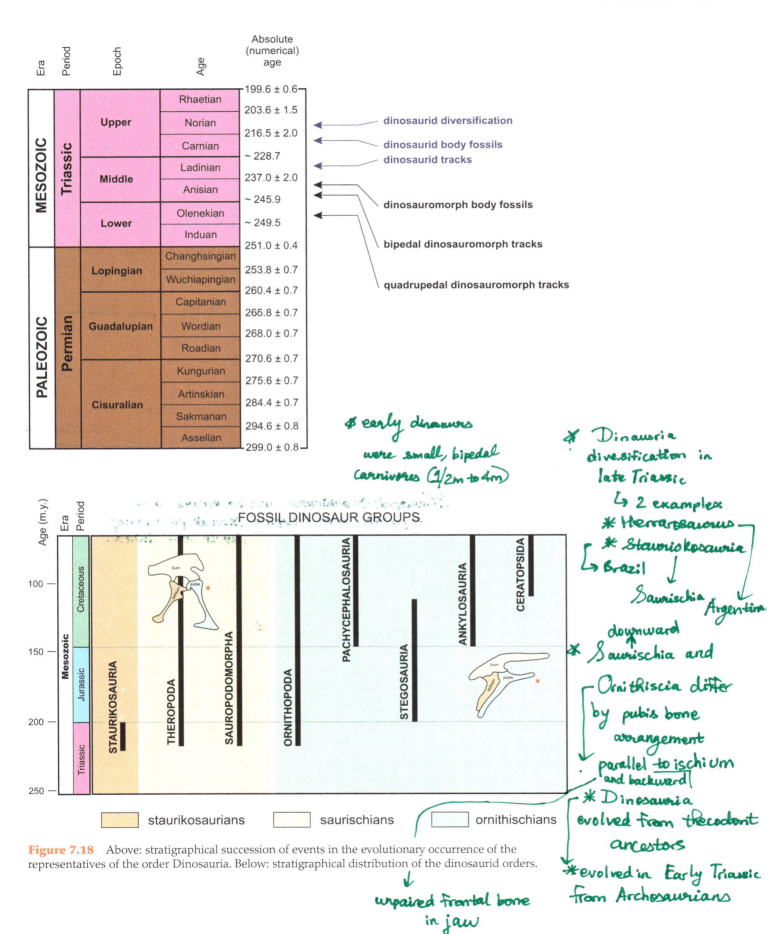

Figure 7.18 Above: stratigraphical succession of events in the evolutionary occurrence of the representatives of the order Dinosauria. Below: stratigraphical distribution of the dinosaurid orders.

Figure 7.19 Example of a representative of the dinosaurid suborder Staurikosauria. Specimen exhibited in the National Museum of Nature and Science, Tokyo; base photograph courtesy Dr. K. Tanaka; published with permission.

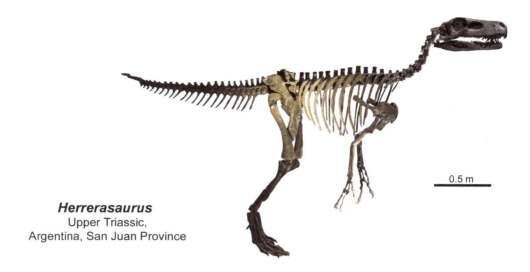

Herrerasaurus
Upper Triassic,
Argentina, San Juan Province

boundary. The other two suborders, namely Theropoda and Sauropodomorpha, flourished in the Jurassic and Cretaceous and became extinct at the Cretaceous/Paleogene boundary.

Suborder Theropoda (Late Triassic-Cretaceous) includes carnivorous dinosaurs of variable size, ranging from about 1 m (e.g., *Microraptor*, etc.) to over 8 m (*Tyrannosaurus*, *Gigantosaurus*, etc.) (Figure 7.20). They were bipedal, capable to run at high speed, which in the case of some species was over 40 km/hour. Larger species, such as *Tyrannosaurus rex* could run at a lower speed. Theropods were equipped with strong jaws and teeth; the backward oriented teeth prevented the prey from escaping and in addition, dentition shows no specialization for chewing. The large-sized heads with an equally developed musculature indicate that these ferocious predators were capable

Figure 7.20 Examples of three representatives of the dinosaurid suborder Theropoda. 1, 3: Specimens from the Senckenberg Natural History Museum, Frankfurt; published with permission. 2: Specimen courtesy of Dr. D.K. Zelenitsky (University of Calgary).

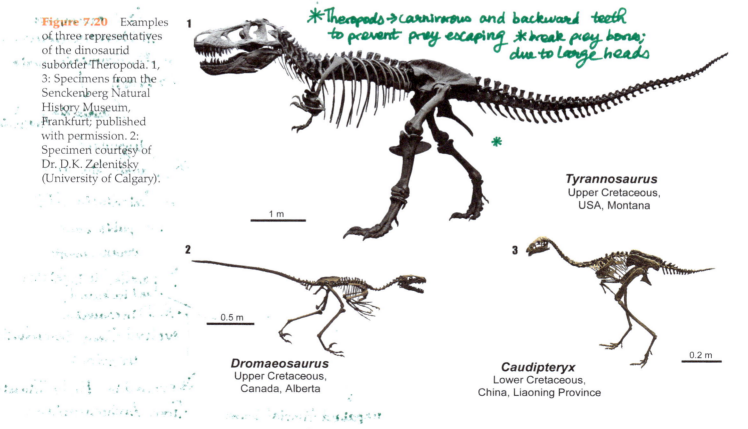

Tyrannosaurus
Upper Cretaceous,
USA, Montana

Dromaeosaurus
Upper Cretaceous,
Canada, Alberta

Caudipteryx
Lower Cretaceous,
China, Liaoning Province

to break the prey bones, not only to slice the prey soft tissues. Eggs of theropods are spectacular occurrences in the fossil record of the group (Figure 7.21).

Sauropodomorphs were the largest terrestrial animals evolved on Earth; some species weighted over 70 tons. The vast majority of these were herbivores but some of the earlier members of this group could have been occasional meat-eaters. There are two groups within the suborder Sauropodomorpha: prosauropods (Late Triassic-Early Jurassic) and sauropods (Jurassic-Cretaceous); the largest sizes were achieved by the sauropods (e.g., *Brontosaurus*, *Apatosaurus*, *Diplodocus*, etc.). Sauropods had massive bodies supported by powerful legs and smaller heads at the end of a long neck. Food digestion was helped with the aid of gastroliths, voluntarily ingested rocks that remained in the reptile stomach for the rest of its life. Sauropods lived in a variety of environments and it is possible some species were migratory.

Ornithischian dinosaurs are more diverse from a morphological point of view when compared to the saurischians as shown by the five suborders included in this

Oviraptor
Upper Cretaceous, Mongolia

Figure 7.21 Examples of theropod dinosaur egg nests. 1: Specimen from the Senckenberg Natural History Museum, Frankfurt; published with permission. 2: Specimen courtesy of Dr. D.K. Zelenitsky (University of Calgary).

Dinosaur egg nest - reconstruction

order. They are characterized by two features: the pubis bone is oriented backward and is close and parallel to the ischium, and present an unpaired bone (*predentary*) in the frontal portion of the jaw.

Suborder Ornithopoda (Late Triassic-Cretaceous) consists of herbivore taxa with the dentition modified for chewing; they occupied the same ecological niches as the modern herbivore mammals. Earlier ornithopods were small-sized, about 1 to 2 m in length and attained larger sizes of about 8 m later in the group's evolution. Ornithopods were bipedal and quadrupedal, the two types of locomotion occurring occasionally in the same species. Back limbs were stronger than the fore limbs and ornithopods were capable of running at a speed of about 15 to 20 km/hour. *Iguanodon*, *Camptosaurus*, *Hadrosaurus*, and *Edmontosaurus* are examples of genera included in this group.

Suborder Stegosauria (Jurassic-Early Cretaceous) are quadrupedal herbivores that could have been as long as 9 m and were probably used for defense. The representatives of this group present *osteoderms* (e.g., plates, spines, and spikes), which are bony structures attached to the skin; larger spines occur in the tail distal portion. Osteoderms present a system of blood vessels and therefore, is inferred that were used for body thermoregulation. *Stegosaurus* is a typical example of genus included in this suborder (Figure 7.22). The group achieved the peak diversity in the Jurassic times and became rare during Early Cretaceous.

Stegosaurus
Upper Jurassic,
USA, Wyoming

Stegosaurus -skull
Upper Jurassic,
unknown location

Figure 7.22 Example of a representative of the dinosaurid suborder Stegosauria. 1: Specimen from the Senckenberg Natural History Museum, Frankfurt; published with permission. 2: Specimen courtesy of Dr. D.K. Zelenitsky (University of Calgary).

Figure 7.23 Example of a representative of the dinosaurid suborder Pachycephalosauria. Specimen courtesy of Dr. D.K. Zelenitsky (University of Calgary).

Suborder Pachycephalosauria (Cretaceous) is characterized by a thickened skull roof. These ornithischians are bipedal and herbivores. *Pachycephalosaurus* is the typical genus of this suborder (Figure 7.23).

Suborder Ankylosauria (Cretaceous) consists of quadrupedal herbivores that had the body covered by bony plates embedded in the skin, resulting in armour-like structure (shield) on the back of the body; the bony armour can be continuous or discontinuous. The tail is thickened at the end and presents a club-like structure, which was used for defense. *Ankylosaurus*, *Euoplocephalus*, and *Saichania* are examples of genera of this suborder (Figure 7.24).

Suborder Ceratopsida (Late Cretaceous) includes quadrupedal and chewing herbivore dinosaurs with the skull protected by a shield that extended backwards to the shoulders; the shield can present anterior horns and could be used for both attack and defense. These dinosaurs could be as long as 10 m. *Triceratops*, *Protoceratops*, and *Chasmosaurs* are examples of ceratopsian dinosaurs (Figure 7.25).

7.12 Birds

Class Aves (Cretaceous-Holocene) evolved in the Late Mesozoic from dinosaurid ancestors, more precisely from theropods. The earliest reptiles with avian features are known in the Late Jurassic (Figure 7.26) but true birds did not evolve until Early Cretaceous.

The flight is often associated with the evolution of feathers; this association is generated by the dominant group of flying vertebrates today, namely the birds. The discovery of an unusual fossil in the Late Jurassic limestones of Solnhofen (Germany) amazed the scientific world, because the small animal presents feathers on the anterior limbs and on the tail; the fossil was named *Archaeopteryx lithographica* (Figure 7.26). This species presents mixed features of reptiles (e.g., tail, anterior members with claws, beak with teeth) and birds (e.g., feathers); the bones of *Archaeopteryx lithographica* have wide hollow spaces, which indicate the distinct possibility that this species could fly. The dominant body features show strong similarities with a number of smaller groups of theropod reptiles. Additional studies showed that it was not capable of avian flight and therefore, the feathers were primarily

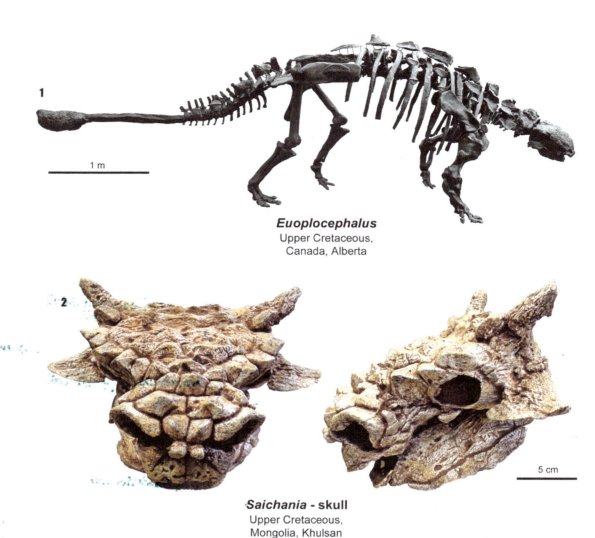

Figure 7.24 Example of representatives of the dinosaurid suborder Ankylosauria. 1: Specimen from the Senckenberg Natural History Museum, Frankfurt. 2: Specimen courtesy of Dr. D.K. Zelenitsky (University of Calgary).

developed to maintain a high body temperature (endothermy) and only secondarily for flight. More recent discoveries of exceptionally well-preserved *Archaeopteryx*-like organisms showed that feathers covered also their legs and trunk. The study of the ancient biotopes in which these animals lived indicated that they were gliders from trees. Despite these advances in understanding the mode of life in the *Archaeopteryx*-like animals, the origins of the avian flight are still unknown. Although the preavian reptiles, namely *Archaeopteryx* and its relatives have significant resemblances with the modern birds, they were not capable of flapped flight.

There are two theories that try to explain the flapped flight origin: arboreal hypothesis and cursorial hypothesis. *Arboreal hypothesis* tries to explain the flight origin from the small-sized non-flying reptiles that glided from the trees. *Cursorial hypothesis* tries to explain the evolution of flight from fast-running theropods, which developed flapped flight on short distances in the attempt to avoid the obstacles at the ground level.

7.12 Birds   193

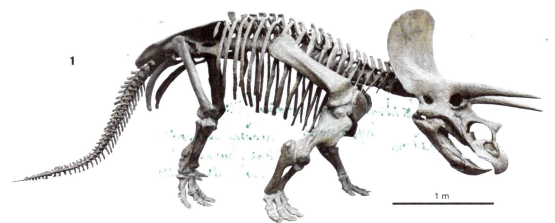

Triceratops
Upper Cretaceous,
USA, Montana

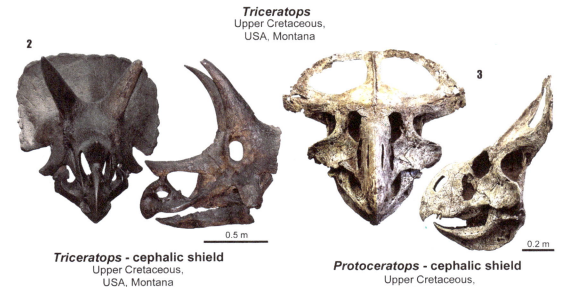

Triceratops - cephalic shield
Upper Cretaceous,
USA, Montana

Protoceratops - cephalic shield
Upper Cretaceous,
Mongolia

Figure 7.25 Example of representatives of the dinosaurid suborder Ceratopsida. 1–2: Specimens from the Senckenberg Natural History Museum, Frankfurt. 3: Specimen courtesy of Dr. D.K. Zelenitsky (University of Calgary).

Archaeopteryx
Upper Jurassic, Solnhofen Limestone,
UE, Germany

Microraptor
Lower Cretaceous,
China, Liaoning Province

Figure 7.26 Examples of theropods along the evolutionary branch that evolved into birds. 1: Specimen exhibited in the Museum of Natural History, Berlin; published with permission. 2: Specimen from the Senckenberg Natural History Museum, Frankfurt; published with permission.

194 Chapter 7 Chordate and Vertebrate Evolution

theropod/saurischia to bird transition in Jurassic/Cretaceous boundary

The morphologic transition from the theropods to the modern bird body plan happened in the proximity of the Jurassic/Cretaceous boundary (Figure 7.27); earliest birds occur in rocks of Cretaceous age (e.g., *Hesperornis*, etc.) (Figure 7.28). Reptile to bird transition involves a variety of morphologic changes in the skeleton, such as: disappearance of teeth, evolution of the pygostyle that supports the

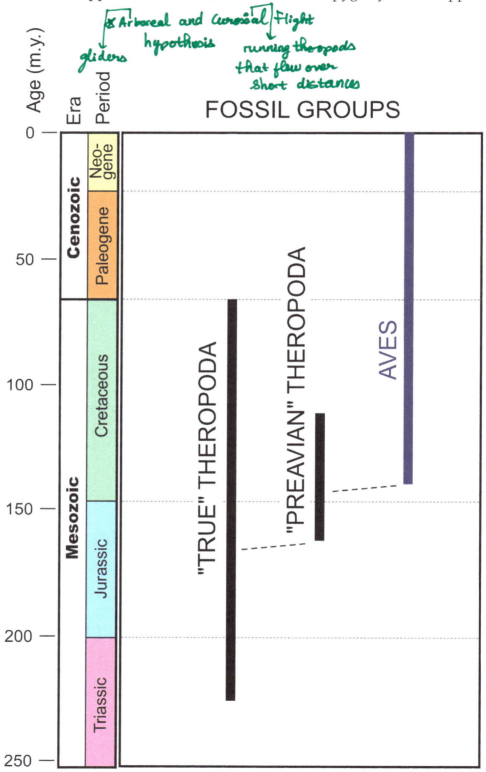

Figure 7.27 Evolution from the theropod reptiles to birds.

Hesperornis

Arboreal and cursorial flight hypothesis → gliders; running theropods that flew over short distances

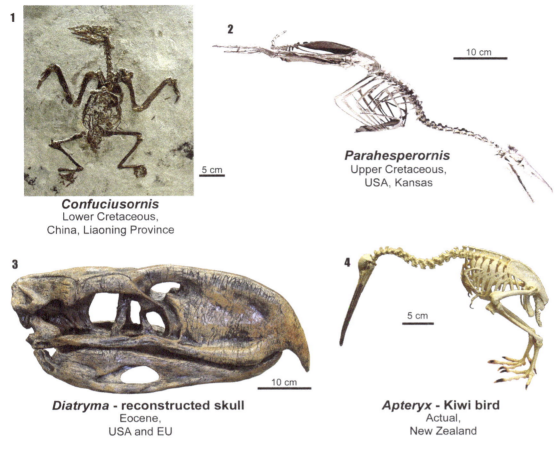

Figure 7.28 Examples of fossil (1–3) and modern (4) birds. 1: Specimen from the Senckenberg Natural History Museum, Frankfurt; published with permission. 2: Specimen exhibited in the National Museum of Nature and Science, Tokyo; base photograph courtesy Dr. K. Tanaka; published with permission. 3–4: Specimens courtesy of Dr. D.K. Zelenitsky (University of Calgary).

tail muscles, toe rearrangement (three to the front and one situated backwards) and loss of the fifth toe, evolution of pneumatic bones, keeled downward projected sternum, loss of some vertebrae, etc.; in addition, in birds feathers cover most of the body surface.

7.13 Mammals

Class Mammalia early history during the Mesozoic times is relatively difficult to study due to the relatively rare well-preserved skeletons. Mammal early history reconstruction is largely based on the skull fragments, isolate body bones, jaws, and teeth.

Transition from the reptilian ancestors to the mammal descendants involves a variety of morphological changes. *Dentary bone* evolved in mammals to completely form the jaw; this contrasts with the reptilian, including therapsids ancestor condition in which the lower jaw consists of several bones; the well-developed dentary increases the chewing efficiency. The other jaw bones became gradually reduced in the mammalian group, and can be found as the three small bones in the middle ear (*malleus*, *incus*, and *stapes*); the three bones became specialized in transmitting the vibrations from the outer ear to the inner ear. *Brain case* is significantly expanded in mammals, whereas in reptiles is small. *Teeth differentiation* is another feature that separates sharply the reptiles and the mammals. Some reptile groups have well-differentiated teeth, which are similar in many respects to those of the mammals. Increasing differentiation in the mammalian teeth is a distinct feature in this group evolution. Teeth replacement in reptiles happens throughout the ontogeny, whereas in mammals they change only once, and in some taxa never throughout their life. Therefore the mammalian dentition appears more efficient as the teeth can adjust during the ontogenetic growth and result in a more precise bite. *Pelvic girdle* is fundamentally different in the two groups. The three bones ilium, ischium, and pubis are distinct in reptiles, and fused in the adult mammals and forming the *coxal bone*. Mammal juveniles have the three bones separated, the fusing process happening relatively late in the ontogenetic development. Other significant changes in the transitions from the reptiles to mammals are the adaptations to preserve the high body temperature and increase the activity during the life time. Mammalian *heart* has four compartments and is similar to that that evolved in birds; mammals present a clear separation between the arterial and venous blood circulation. Mammal body is covered with *hair* rather than scales as in the reptiles; however, it is possible that some therapsid reptiles living in colder climates to have developed hair. Evolution of a *diaphragm*, which separates the thoracic and abdominal portions of the body cavity in the mammalian group, increases the efficiency of the oxygen intake. Reptile *reproduction* is through eggs; some species protect their eggs. Evolved mammals give birth to one or a small number of juveniles, which are nurtured. Primitive mammals, such as the monotremes lay eggs, reproduction mechanism inherited from the reptilian ancestors. Modern mammals are subdivided into three groups according to the reproduction strategy: monotremes, marsupials, and placentals. *Monotremes* lay eggs. *Marsupials* give birth to living juveniles, which are nurtured in a pouch. *Placentals* give birth to living juveniles, and feed them with milk produced by the mammary glands.

The earliest mammals (Late Triassic-Jurassic) were possibly monotremes; in general they were small-sized insectivores (Figure 7.29). Modern monotremes present significant similarities with the Late Triassic-Jurassic mammals especially in the skull bone features. *Multituberculates* of the Jurassic-Paleogene are small-sized rodent-like mammals that dominated the Mesozoic mammal faunas. Marsupials evolved during the Cretaceous but did not diversify until Paleogene. The earliest occurrence of the placentals is in the proximity of the Middle/Late Jurassic boundary.

The mammal Cenozoic fossil record is more complete when compared with that of the Mesozoic representatives of this class. There are two distinct phases in the mammalian evolution during the Cenozoic. Marsupials dominated in the Early Cenozoic and placentals in the Late Cenozoic; both groups evolved in the Mesozoic, during the Cretaceous and Jurassic, respectively. Besides the differences in the reproduction mechanisms between marsupials and placentals there are other significant differences between the two groups among which is mentioned the dentition replacement that is complete in placentals and restricted to the premolars in the posterior part of the jaw in the marsupials.

Marsupial evolutionary occurrence and early evolution happened in North America, and during the Cretaceous most of the species were restricted to the North American continent. Group migration resulted in the colonization of the other continents. Marsupials are dominantly vegetarians; among the carnivorous representatives of the group are the saber-tooth-like *Thylacosmilus* of the Pliocene

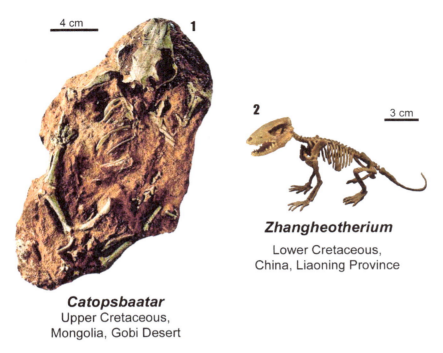

Figure 7.29 Examples of Mesozoic mammals. 1: Specimen by Hurum and Kielan-Jaworowska (2008, fig. 1: A1); © Acta Palaeontologica Polonica; published with permission. 2: Specimen exhibited in the National Museum of Nature and Science, Tokyo; base photograph courtesy Dr. K. Tanaka; published with permission.

and dog-like *Borhyaena* of the Miocene. Modern marsupials include kangaroos, wallabies, possums, koalas, and wombats. Most of the living taxa of marsupials live in Australia.

McKenna and Bell (1997) subdivided the placental mammals into five groups of orders: *Archonta*, *Insectivora*, *Anagalida*, *Creodonta* and *Carnivora*, and *Ungulata* (Figure 7.30), which are not necessarily evolutionary taxonomic units. *Archonta* includes the placentals with relatively small morphologic differences when compared to the basal placentals. Bats (order Chiroptera) and primates (order Primata) are parts of this group. The former is the only mammalian group that evolved flapped flight and the latter led to the evolution of bipedal, highly intelligent species. Bats evolved in the Paleocene and primates in the Latest Cretaceous. *Insectivora* is a group of small-sized placentals, which evolved during the Latest Cretaceous; moles are the best known representatives of this group. *Anagalida*, which is also known as *Glires*, includes among others the rodents (order Rodentia) and lagomorphs (order Lagomorpha); rats and mice are typical representatives of the rodents, whereas the rabbits and hares are typical lagomorphs. The group evolved in the Paleocene and it is sparsely represented in the fossil record during the Paleocene and Eocene; its diversification began in the proximity of the Eocene/Oligocene boundary, when both rodents and lagomorphs became morphologically well-separated from their ancestors. *Creodonts and carnivores* are also known as *Ferae*. Order Creodonta consists of carnivorous placentals and is known only as fossils from the Paleogene-Neogene stratigraphical interval. Carnivores are morphologically and taxonomically more diverse than the creodonts; felines are examples of carnivores, and in the Neogene times evolved ferocious predators (Figure 7.31). *Ungulata* is the largest and most

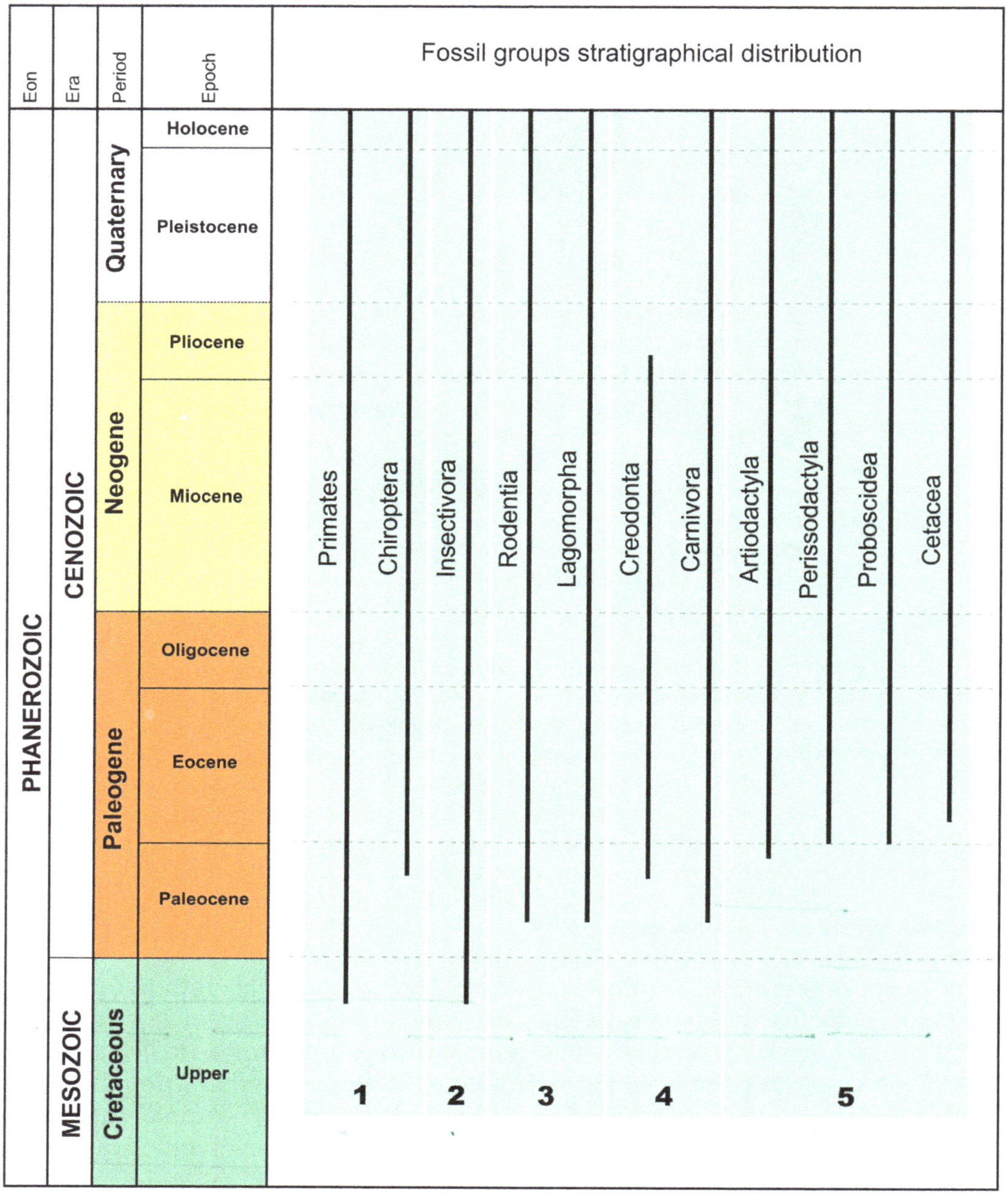

Figure 7.30 Stratigraphical distribution of the groups of placental mammal orders recognized by McKenna and Bell (1997). 1: Archonta, 2: Insectivora, 3: Anagalida, 4: Creodonta and Carnivora, and 5: Ungulata.

Figure 7.31 Examples of a Pleistocene carnivore. 1: Specimen from the Senckenberg Natural History Museum, Frankfurt; published with permission. 2: Specimen courtesy of Dr. D.K. Zelenitsky (University of Calgary).

Smilodon
Pleistocene,
USA, California

diverse group of placental orders. This group includes among others order Artiodactyla (deer, antelopes, camels, lamas, etc.), order Perissodactyla (horses, rhinoceroses, etc.), order Proboscidea (elephants, mastodons, etc.), and order Cetacea (whales, dolphins, etc.).

CHAPTER CONCLUSIONS

- Chordate animals are included in three phyla: Cephalochordata, Conodonta, and Tunicata.
- Earliest chordates are evolved in the Early Cambrian.
- Best fossil record among chordates is that of the phylum Conodonta due to evolution of a chewing apparatus consisting of apatite and organic matter.
- Early Cambrian *Haikouichthys* is the earliest agnathan in the history of life.
- Most of the agnathan taxa are recorded in the Ordovician-Devonian stratigraphical interval.
- Acanthodii are the earliest gnathostomes in the history of life.
- Acanthodians present cartilaginous skeletons with the only bony components as anterior spines in the fins.
- Most of the placoderm species are included within orders Antiarchi and Arthrodira.
- Placoderms evolved the largest predators in the Devonian times, such as *Dunkleosteus* and *Tityosteus*.
- Placoderms became extinct at the Devonian/Carboniferous boundary.
- Chondrichthyes are also known as cartilaginous fishes and were major predators during the Carboniferous-Holocene times.
- Neogene Chondrichthyes genus *Carcharodon* is one of the largest marine predators in the history of life.
- Bony fishes are included within class Osteichthyes, which is subdivided into subclasses Actinopterygii and Sarcopterygii according to the fin morphology.
- Subclass Actinopterygii became extremely diverse after the Devonian/Carboniferous boundary and represents the most diverse group of living fishes; it is also the only fish group that evolved flight.
- Rhipidistian sarcopterygians evolved into land tetrapods in the Late Devonian times.
- Rhipidistian genus *Eusthenopteron* is most likely the direct ancestor of earliest land vertebrates.
- Representatives of the amphibian order Temnospondyli are the ancestors of modern amphibians included in the lissamphibian group.
- Reptile ancestors are among the amphibian order Anthracosauria.
- Earliest reptiles evolved in the Early Carboniferous and were diapsids; *Petrolacosaurus* is the oldest known reptile in the fossil record.
- Pelycosaurs were synapsid reptiles that evolved the earliest mechanism of body thermoregulation among reptiles; they dominated the terrestrial environments during the Late Carboniferous-Early Permian times.

- Representatives of the reptile order Therapsida evolved in the Early Permian and dominated the terrestrial environments started in the Late Permian; in the Late Triassic this group gave birth to class Mammalia.
- Reptiles adapted to aquatic environments several times during their evolution; ichthyosaurids and sauropterygians were among the most ferocious predators in the Mesozoic seas and oceans.
- Flying reptiles of the Mesozoic are included within order Pterosauria; representatives of this order evolved iteratively several times during the Late Triassic-Cretaceous times.
- Dinosaurs were terrestrial reptiles that evolved gigantic sizes in the Jurassic and Cretaceous times.
- Theropod dinosaurs evolved into birds in the proximity of the Jurassic/Cretaceous boundary; *Archaeopteryx lithographica* is the first discovered feathered reptile with mixed features, reptilian and avian.
- Earliest birds occur in the Early Cretaceous times.
- Mammals were small-sized throughout the Mesozoic times.
- Mammal diversification began in the Paleocene times.
- Primates evolved in the Latest Cretaceous times.

CHAPTER 8

PLANT EVOLUTION

CONTENT

8.1 Colonization of Terrestrial Environments
8.2 Rhyniophytes
8.3 Lycophytes, Sphenophytes, and Pteridophytes
8.4 Spermatophytes

Chapter Conclusions

8.1 Colonization of Terrestrial Environments

Evolution of kingdom Plantae or *green plants* began with the colonization of land by aquatic organisms and there are many unknowns on how this process happened. The beginnings of land colonization can be recognized in the fossil record by the occurrence of spores, which are often described briefly as reproductive structures of lower plants; spores are highly resistant during fossilization and therefore, fossilize often. Spores are produced by several groups of land plants and among them by the representatives of *division Bryophyta*, the land plants with the simplest morphology in the modern floras; this division includes hornworts, liverworts and mosses. These organisms do not have a specialized system of nutrient transport through the plant body, and the substances are transported through impregnation and osmosis or, more rarely through a tubular structure; they are non-vascular plants and *Agalophyton* is such a genus of Early Devonian age (Edwards, 1986). The fossil record of the division Bryophyta is sparse and the oldest fossils of this division are of Early Devonian age. Spores are also produced by vascular plants, which are known in the fossil record starting in the Middle Silurian times. Evolution of vascular tissue, such as *xylem* and *phloem* marks the transition to *division Tracheophyta*, which includes all vascular plants.

The fossil record of the spores begins in the Middle Ordovician. Therefore, it is possible that the spores recognized in the Middle Ordovician-Lower Silurian stratigraphical interval belong to a different group. Notably, representatives of the kingdom Fungi can produce spores as reproductive structures, and this makes the representatives of this kingdom the most reliable candidates for the green plant ancestry. One distinct possibility is that the earliest land plants evolved from the fungi washed out on the beaches during the Early Paleozoic times. Notably, traces of arthropods are reported from these environments from throughout the Cambrian-Silurian stratigraphical interval. Therefore, it seems reasonable to infer that the washed out fungi and algae represented the base of true trophic chains in the Early Paleozoic beaches environments. How exactly such fungi and possibly algae evolved into the earliest land plants is unknown.

8.2 Rhyniophytes

Class Rhyniophyta (Middle Silurian-Devonian) includes the earliest land plants in the fossil record; they are vascular plants. Seemingly *Cooksonia*'s oldest occurrence is in the Middle Silurian (Edwards and Freehan, 1980); it is a small plant with a height <10 cm, which has a slender stem that can be branched dichotomously and bears the reproductive organs termed *sporangia* at the terminal end of the branches. One very simple conducting strand occurs occasionally in the representatives of this genus (Edwards and others, 1992).

This class diversified in the Devonian times and such diversification is associated with the gradual evolution of the conducting strands, woody tissue, and leaves (Figure 8.1). Genus *Rhynia* from the Early Devonian is a small-sized plant with an average height of 5 cm, which was attached to the substratum with a delicate rhizome; a section through the stalk indicates the occurrence of one conducting strand. In the Middle Devonian times more evolved rhyniophytes occur and for example, *Hyenia* could be as high as 20 cm, with small but distinct leaves along the stalk. The woody tissue is further developed in plants like *Cladoxylon* of Middle Devonian-Early Carboniferous age; *Cladoxylon* was probably over half meter in

Figure 8.1 Examples of representatives of the class Rhyniophyta. All specimens from the collections of the Museum of Natural History, Berlin; published with permission.

height and the well-developed branching stalks with woody aspect seemingly indicate that it could colonize terrestrial environments further away from the water pools and rivers.

8.3 Lycophytes, Sphenophytes, and Pteridophytes

Three classes of spore-producing plants are morphologically advanced when compared with the primitive and extinct class Rhyniophyta. The three classes are: Lycophyta, Sphenophyta, and Pteridophyta; all of them were diverse in the earlier part of the stratigraphical ranges that began in the Silurian-Devonian and exist as minor components of the modern floras.

Class Lycophyta (Late Silurian-Holocene) evolved leaves and the occurrence of these structures is apparent in the earliest lycophyte taxa of the Late Silurian. For example, the genus *Baragwanathia* of the Late Silurian presents long elongate leaves along the stem (Figure 8.2). Another interesting and characteristic morphological feature is represented by the position of the reproductive structures (sporangia), which are no longer situated at the terminal end of the branches as in rhyniophytes,

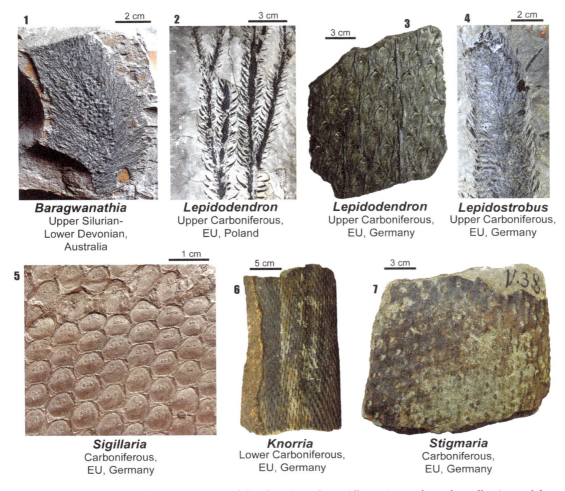

Figure 8.2 Examples of representatives of the class Lycophyta. All specimens from the collections of the Museum of Natural History, Berlin; published with permission.

but are migrated toward the junction between leaves and stem. In some lycophyte taxa sporangia evolved in distinct reproductive structures on the lycophyte plant and in the past some were described as distinct genera. When compared with the rhyniophytes, lycophytes also have well-developed roots. Lycophytes occur in the modern floras as smaller plants known informally as *club mosses*. But the fossil record of the lycophytes shows that the representatives of this class evolved large-sized trees during the Carboniferous times, when they contributed massively to the formation of vast amounts of coals. *Lepidodendron* is a typical example of large-sized lycophyte trees, which occurs as tree stems, but also reproductive structures that are described as a distinct genus: *Lepidostrobus* (Figure 8.2). *Sigillaria* and *Knorria* are other examples lycophyte large-sized trees of Carboniferous age. The root structures of the lycophyte trees are formalized as distinct genera and *Stigmaria* represents one such example (Figure 8.2).

Class Sphenophyta (Late Devonian-Holocene) is represented in the modern floras by horsetails; the only living genus is *Equisetum*. The representatives of this class are characterized by segmented stem that presents a hollow interior with tubular shape; leaves are arranged in verticiles at the joint between two adjacent segments of the stem. Sphenophyte genera were described for different parts of the plant: stem (e.g., *Calamites*), leaves (e.g., *Annularia, Asterophyllites*), and reproductive structures (e.g., *Macrostachya*) (Figure 8.3).

Figure 8.3 Examples of representatives of the class Sphenophyta. All specimens from the collections of the Museum of Natural History, Berlin; published with permission.

Figure 8.4 Examples of representatives of the class Pteridophyta. Both specimens from the collections of the Museum of Natural History, Berlin; published with permission.

Class Pteridophyta (Middle Devonian-Holocene) includes the ferns, which occur in the modern floras and are abundant in certain regions; they are the most abundant spore-bearing plants in the living floras. The stem grows underground and from it fronds emerge; sporangia are situated on the lower side of the fronds. Most of the fern genera described from the fossil record are based on frond morphology (e.g., *Oligocarpia*, *Pecopteris*, etc.) (Figure 8.4). Pteridophytes diversified in the Carboniferous times and were reduced in the Permian with the growing diversity of the gymnosperms.

8.4 Spermatophytes

The next major leap in the evolution of the land plants is represented by the evolution of seeds, which are structures developed in the course of plant reproduction. These land plants are spermatophytes, which are informally referred to as vascular seeded plants, and included in three classes: Pteridospermatophyta, Gymnospermatophyta, and Angiospermatophyta. Of them, Gymnospermatophyta and Angiospermatophyta are the dominant classes in the living floras. The seeds can be completely enclosed within a protective structure or left exposed (Figure 8.5).

Class Pteridospermatophyta (Late Devonian-Jurassic, Early Cretaceous) includes land plants with a general fern aspect that reproduce through seeds rather than spores. The representatives of this class are also known as *seed ferns*. Pteridospermatophytes are rare in the Late Devonian and reached the diversity peak during the Carboniferous-Permian times. *Alethopteris*, *Lonchopteris*, *Mariopteris*, and *Neuropteris* are examples of pteridospermatophyte genera (Figure 8.6). After the Permian/Triassic major crisis in the history of life, pteridospermatophytes entered a period of decline, and eventually became extinct in the proximity of the Jurassic/Cretaceous boundary.

Class Gymnospermatophyta (Carboniferous-Holocene) became dominant in the Permian times and this setting remained unchanged for most of the Mesozoic. Morphologically the gymnosperms are extremely diverse and are subdivided into

Figure 8.5 Reproductive structures of the classes Pteridospermatophyta (1), Gymnospermatophyta (2) and Angiospermatophyta (3). 1: specimen from the collections of the Museum of Natural History, Berlin; published with permission. 2–3: specimens courtesy of Dr. L.V. Hills (University of Calgary); published with permission.

Figure 8.6 Examples of representatives of the class Pteridospermatophyta. All specimens from the collections of the Museum of Natural History, Berlin; published with permission.

several orders; among them are Coniferales, Bennettitales, and Ginkgoales. *Order Coniferales* (Carboniferous-Holocene) is the most diverse order of this class. The representatives of this order are the dominant gymnosperms in the modern floras; *Walchia*, *Ullmannia*, and *Pseudovoltzia* are examples of fossil conifer genera (Figure 8.7). *Order Bennettitales* (Triassic-Cretaceous) is the only extinct gymnosperm order and its representatives are relatively rare as fossils; one example of genus of this order is *Nilssonia* (Figure 8.7). *Order Ginkgoales* (Late Triassic-Holocene) is relatively rare both in the fossil record and living floras; *Baiera*, *Ginkgoidium*, and *Ginkgoites* are examples of genera included in this order (Figure 8.7).

Class Angiospermatophyta (Cretaceous-Holocene) includes the most evolved plants on Earth, and present as characteristic feature the seeds enclosed in a fruit. They are known informally as *flowering plants*. Angiosperms became dominant plants on Earth in the Latest Cretaceous and this setting is encountered today; it is estimated that angiosperms form more than 90% of the living species of land plants.

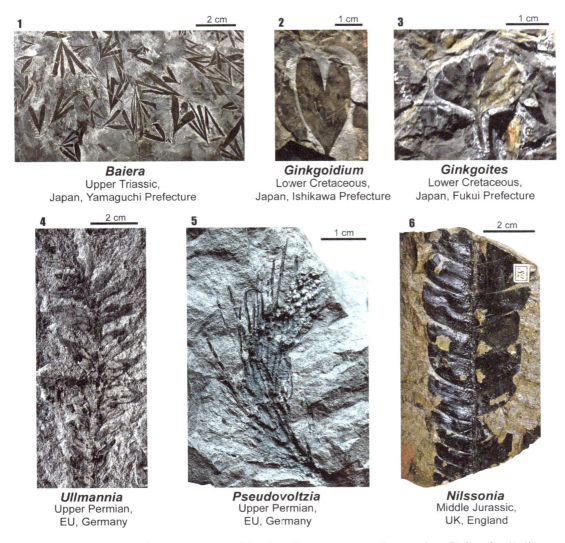

Figure 8.7 Examples of representatives of the class Gymnospermatophyta, orders Ginkgoales (1–3), Coniferales (4–5) and Bennettitales (6). 1–3: Specimens exhibited in the National Museum of Nature and Science, Tokyo; base photograph courtesy Dr. K. Tanaka; published with permission. 4–6: Specimens from the collections of the Museum of Natural History, Berlin; published with permission.

CHAPTER CONCLUSIONS

- Earliest spores in the fossil record are of Middle Ordovician age.
- Earliest land plants occur in the Middle Silurian times.
- Rhyniophytes are the earliest vascular land plants; representatives of this group present a gradual increase in height and degree of development of the woody tissue.
- Rhyniophytes reproduction was through spores.
- Lycophytes, sphenophytes, and pteridophytes are the evolved groups of plants in which reproduction is through spores; all evolved in the Late Silurian-Devonian times and exist in the modern flora.
- Spermatophytes include all the seed plants.
- Gymnospermatophytes dominated the land floras during the Permian-Cretaceous times.
- Angiospermatophytes are dominant land plants today.

REFERENCES

Baliński, A. "First colour-patterned strophomenide brachiopod from the earliest Devonian of Podolia, Ukraine," *Acta Palaeontologica Polonica*, 55 (2010): 695–700.

Brain, C.K., Prave, A.R., Hoffmann, K.-H., Fallick, A.E., Botha, A., Herd, D.A., Sturrock, C., Young, I., Condon, D.J., and Allison, S.G. "The first animals: ca. 760-million-year-old sponge-like fossils from Namibia," *South African Journal of Science*, 108 (2012): 1–8.

Chen, J.-Y., Huang, D.-Y., Peng, Q.-Q., Chi, H.-M., Wang, X.-Q., and Feng, M. "The first tunicate from the Early Cambrian of South China." *Proceedings of the National Academy of Sciences of the United States of America*, 100 (2003): 8314–8318.

Darwin, C. *On the Origin of Species by Means of Natural Selection, or the Preservation of Favoured Races in the Struggle for Life* (London: John Murray, 1859), 502 p.

Dlussky, G.M., and Radchenko, A.G. "Two new primitive ant genera from the late Eocene European ambers," *Acta Palaeontologica Polonica*, 54 (2009): 435–441.

Dzik, J., "*Yunnanozoon* and the ancestry of chordates," *Acta Palaeontologica Polonica*, 40 (1995): 341–360.

Edwards, D.S. "*Agalophyton major*, a non-vascular land-plant from the Devonian Rhynie Chert," *Botanical Journal of the Linnaean Society*, 93 (1986): 173–204.

Edwards, D., and Feehan, J. "Records of *Cooksonia*-type sporangia from late Wenlock strata in Ireland," *Nature*, 287 (1980): 41–42.

Edwards, D., Davies, K.L., and Axe, L. "A vascular conducting strand in the early land plant *Cooksonia*," *Nature*, 357 (1992): 683–685.

Ehrenberg, C.G. "Verbreitung und Einfluss des mickroskopischen Lebens in Süd- und Nord-Amerika," *Abhandlungen der Königlichen Akademie der Wissenschaften zu Berlin* (1841): 291–446. [published in 1843]

Ehrenberg, C.G. *Mikrogeologie* (Leipzig: L. Voss, 1854), 374 p.

Encelius, C. *De Re Metallica, Hoc Est, de Origine, Varietate, & Natura Corporum Metallicorum, Lapidum, Gemmarum, atq ; aliarum, que ex fodinis, eruuntur, rerum, ad Medicine usum deservientium, Libri III* (Frankfurt: Chr. Egenolphum, 1551), 271 p.

Erben, H.K. "Über den Ursprung der Ammonoidea," *Biological Reviews*, 41 (1966): 641–658.

Fedonkin, M.A., Vickers-Rich, P., Swalla, B.J., Trusler, P., and Hall, M. "A new metazoan from the Vendian of the White Sea, Russia, with possible affinities to the ascidians," *Paleontological Journal*, 46 (2012): 1–11.

Georgescu, M.D. "Upper Jurassic–Cretaceous planktonic biofacies succession and the evolution of the Western Black Sea Basi," in *Regional and petroleum geology of the Black Sea and surrounding region*, ed. A.G. Robinson, AAPG Memoir 68, 1997, p. 169–182.

Georgescu, M.D. "Microfaunal abundance fluctuations in the Western Black Sea (Romanian offshore, Cretaceous to Pliocene)," in *Micropaleontologic proxies for sea-level change and stratigraphic discontinuities*, eds H. Olson and R.M. Leckie (SEPM Special Publication 75, 2003), p. 301–315.

Georgescu, M.D. "Upper Albian-lower Turonian non-schackoinid planktic foraminifera with elongate chambers: morphology reevaluation, taxonomy and evolutionary classification," *Revista Española de Micropaleontología*, 41 (2009a): 255–293.

Georgescu, M.D. "On the origins of Superfamily Heterohelicacea Cushman, 1927 and the polyphyletic nature of plantic foraminifera," *Revista Española de Micropaleontología*, 41 (2009b): 107–144.

Georgescu, M.D. "Origin, taxonomic revision and evolutionary classification of the late Coniacian-early Campanian (Late Cretaceous) planktic foraminifera with multichamber growth in the adult stage," *Revista Española de Micropaleontología*, 42 (2010): 59–118.

Georgescu, M.D. "Evolutionary classification and nomenclature of the Cretaceous planktic foraminifera with the chambers alternately added with respect to the test growth axis," in *Evolutionary classification and english-based nomenclature in cretaceous planktic foraminifera*, eds M. D. Georgescu and C. M. Henderson (New York: Nova Science Publishers, 2014), p. 129–248.

Georgescu, M.D. *Microfossils through time: An introduction* (Stuttgart: Schweizerbart, 2018), 400 p.

Georgescu, M.D. and Braun, W.K. "Devonian Charophyta of Western Canada," *Revista Española de Micropaleontología*, 38 (2006a): 1–9.

Georgescu, M.D. and Braun, W.K. "Jurassic and Cretaceous Charophyta of Western Canada," *Micropaleontology*, 52 (2006b): 357–369.

Georgescu, M.D. and Braun, W.K., 2013. "Volgian foraminifera and biostratigraohy of the Husky Formation, Arctic Slope region of northwestern Canada," in *Foraminifera. Aspects of classification, stratigraphy, ecology and evolution*, ed. M. D. Georgescu (New York: Nova Science Publishers, 2013), p. 21–58.

Georgescu, M.D. and Henderson, C.M. *Geological history of life. A manual for non-major students* (Dubuque: Kendall Hunt Publishing Company, 2013), 163 p.

Georgescu, M.D. and Huber, B.T. "Early evolution of the Cretaceous serial planktic foraminifera (late Albian-Cenomanian)," *Journal of Foraminiferal Research*, 39 (2009): 335–360.

Gesner, C., 1565. "De Rerum Fossilium, Lapidum et Gemmarum maximé, figuris & similitudinibus," in *De Omni Rerum Fossilium Genere, Gemmis, Lapidibus, Metallis, et Huiusmodi, Libri Aliquot, Plerique Nunc Primum Editi* (Tiguri: Jacob Gesner, 1565), p. 1–169.

Hamilton, H.C. and Falconer, W. *The geography of Strabo* (London: Henry G. Bohn, 1854), 410 p.

Hou, X. and Bergström, J. "The Chengjiang fauna-the oldest preserved animal community," *Paleontological Research*, 7 (2003): 55–70.

Hurum, J.H. and Kielan-Jaworowska, Z. "Postcranial skeleton of a Cretaceous multituberculate mammal *Catopsbaatar*," *Acta Palaeontologica Polonica*, 53 (2008): 545–566.

Israelsky, M.C. "Oscillation chart," *Bulletin of the American Association of Petroleum Geologists*, 33 (1949): 92–98.

Luterbacher, H. "Foraminifera from the Lower Cretaceous and Upper Jurassic of the northwestern Atlantic," in *Initial Reports of the Deep Sea Drilling Project Leg 11*, ed. A. G. Kaneps (Washington, DC: United States Government Printing Office, 1972), p. 561–593.

Johnson, J.H., 1961. Fossil algae from Eniwetok, Funafuti and Kita-Daitō-Jima. *United States Geological Survey Professional Paper*, 260-Z, p. 907–950.

MacMahon, J.H. *The refutation of all heresies by Hippolytus* (Edinburgh: T. & T. Clark, 1868), 508 p.

Ogg, J.G., Ogg, G., and Gradstein, F.M., 2008. *A concise geological time scale* (Cambridge: Cambridge University Press, 2008), 177 p.

Pease, A.S. "Fossil fishes again," *Isis*, 33 (1942), p. 689–690.

Sagan, C. *Cosmos* (New York: Ballantine Books, 1985), 324 p.

Schopf, J.W. *Cradle of life. The discovery of Earth's earliest fossils* (Princeton: Princeton University Press, 1999), p. 1–367.

Shu, D.-G., Chen, L., Han, J., and Zhang, X.-L. "An Early Cambrian tunicate from China," *Nature*, 441 (2001): 472–473.

Shu, D.G., Conway Morris, S., Han, J., Zhang, Z.-F., Yasui, K., Janvier, P., Chen, L., Zhang, X.-L., Liu, J.-N., Li, Y., and Liu, H.-Q. "Head and backbone of the Early Cambrian vertebrate *Haikouichthys*," *Nature*, 421 (2003): 526–529.

Simpson, G.G., 1963. *Principles of Animal Taxonomy*. Columbia University Press, 247 pp.

Skovsted, C.B. and Peel, J.S. "Small shelly fossils from the argillaceous facies of the Lower Cambrian Forteau Formation of western Newfoundland," *Acta Palaeontologica Polonica*, 52 (2007): 729–748.

Stehli, F.G., Creath, W.B. "Foraminiferal ratios and regional environments," *The American Association of Petroleum Geologists Bulletin*, 48 (1964): 1810–1827.

Stradner, H. and Allram, F. "Notes on an enigmatic siliceous cyst, Middle America Trench, Deep Sea Drilling Project Hole 490," in *Initial Reports of the Deep Sea Drilling Project Leg 66*, ed. M. Lee (Washington, DC: United States Government Printing Office, 1982), p. 641–642.

Tighe, M.B. and Gurney, H. *The Works of Apuleius Comprising the Metamorphoses, or Golden Ass, the God of Socrates, the Florida and his Defence, or a Discourse on Magic* (London: George Bell and Sons, 1878), 533 p.

Tomida, S. "Coleoid *Spirulirostra* (Cephalopoda, Mollusca) from the Miocene Mizunami Group, Central Japan," *Transactions and Proceedings of the Palaeontological Society of Japan*, 168 (1992): 1329–1338.

Vincent, E., Lehmann, R., Sliter, W.V., and Westberg, M.J. "Calpionellids from the Upper Jurassic and Neocomian of Deep Sea Drilling Project Site 416, Moroccan Basin, Eastern North Atlantic," in *Initial Reports of the Deep Sea Drilling Project Leg 50*, eds L. N. Stout, and P. Worstell (Washington, DC: United States Government Printing Office, 1980), p. 439–465.

Zhu, M., Zhao, W., Jia, L., Lu, J., Qiao, T., and Qu, Q. "The oldest articulated osteichthyan reveals mosaic gnathostomes characters," *Nature*, 458 (2009): 469–474.

INDEX OF GENERIC NAMES

A

Acaciella, 98
Acanthodes, 172, 173
Acanthoscaphites, 145, 146
Acanthostega, 179
Acanthoteuthis, 3
Aclistochara, 33
Acrothele, 126
Actinocamax, 150, 151
Actinosepia, 150, 152
Actinostroma, 116
Adelphoceras, 141
Aechmina, 156
Aequitriradites, 36
Agalophyton, 201
Agelas, 39
Ainoceras, 145, 147
Aldanella, 133
Alethopteris, 17, 205, 206
Aletograptus, 164
Alievium, 38
Amaltheus, 9, 147, 148
Ammobaculites, 58
Ammomarginulina, 58
Anabaena, 32
Anadara, 136, 137
Anarcestes, 144, 145
Anaticinella, 69
Anatrypa, 130
Ancyloceras, 145, 146
Andrias, 181
Animikiea, 98
Ankylosaurus, 191
Annularia, 204
Anomalocaris, 111
Apateon, 179
Apatosaurus, 189
Apteryx, 195
Arca, 136, 137
Archaeoguembelitria, 84
Archaeopteryx, 191, 192, 193, 199
Archaeoscillatoriopsis, 96
Archaeotrichion, 96

Archimedes, 123
Argonauta, 151, 152
Asterophyllites, 35, 204
Asteropyge, 155
Astrangia, 120
Astraspis, 170
Astrhelia, 120
Ataxioceras, 147, 149
Athleta, 135
Athyris, 130
Atractosella, 121
Atrypa, 130
Aulacoceras, 144
Aysheaia, 111

B

Bactrites, 141, 144
Baiera, 207
Balanus, 45
Bangiomorpha, 102, 103
Baragwanathia, 203
Basidechenella, 155
Belemnitella, 150, 151
Belemnoteuthis, 150, 151
Beyrichia, 156
Borhyaena, 197
Bothriolepis, 173, 174
Branchiostoma, 169
Brontosaurus, 189
Bulbophragmium, 58

C

Cactocrinus, 20
Cadomites, 147, 148
Calamites, 204
Calpionella, 38
Cambropallus, 153, 154
Camptosaurus, 190
Capitosaurus, 179, 180
Carcharodon, 7, 175
Cardiopteris, 19

Cardium, 7
Caryosphaeroides, 102
Catenipora, 117
Catopsbaatar, 197
Caudipteryx, 188
Cephalaspis, 171
Ceramium, 32
Ceratodus, 178
Ceraurus, 155
Cerithium, 135
Chama, 139
Chara, 32, 33
Charnia, 104
Chasmosaurs, 191
Chelonia, 183
Cheungkongella, 169
Chirognathus, 49
Chiton, 132
Chuaria, 103
Cidaris, 162
Cladoselache, 175
Cladoxylon, 202
Clavulinoides, 58
Cloudina, 105, 106, 113, 117
Clydonautilus, 142, 143
Cnemidopyge, 154
Coeloma, 156
Coeloptychium, 116
Coelurosauravus, 183
Composita, 130
Coccosteus, 173, 174
Conchorhynchus, 149
Confuciusornis, 195
Conocaryomma, 38
Conocoryphe, 154
Conopeum, 122
Conotubus, 106, 117
Constellaria, 122
Cooksonia, 202
Coscinoasterias, 162
Cranaena, 131
Crassostrea, 136, 138
Crisia, 122
Cruziana, 3

Cucullaea, 136, 137
Cupularostrum, 128
Cyathidites, 33
Cybister, 18
Cycleryon, 156
Cyclomedusa, 104
Cylicioscapha, 134
Cynognathus, 56
Cyrtograptus, 165
Cyrtopleura, 139, 140
Cytherelloidea, 156

D

Dactylioceras, 147, 148
Deltoblastus, 160, 161
Daphnogene, 18
Dendrocystites, 160
Dendrophyllia, 120
Dentalium, 132
Derbyia, 127
Diatryma, 195
Dickinsonia, 104
Didymograptus, 164, 165
Dieneroceras, 145, 146
Dimetrodon, 182
Diplacanthus, 172, 173
Diplodocus, 189
Diplomystus, 176
Dipterus, 178
Discoceratites, 145, 146
Dosinia, 66, 139
Drepanaspis, 171
Dromaeosaurus, 188
Dunkleosteus, 173, 174, 199

E

Eatonia, 128
Echmatocrinus, 112
Edaphosaurus, 182
Ediacaria, 104
Edmontosaurus, 190
Elrathia, 154
Encope, 163
Endoceras, 9
Entosphaeroides, 98
Eoleptonema, 96
Eoastrion, 98
Eosphaera, 98

Eopelobates, 181
Eospirifer, 129
Eotetrahedrion, 102
Epipolysemia, 181
Equisetum, 204
Eretmocrinus, 161
Erugatocyathus, 117
Eryops, 179
Eubostrycoceras, 145, 147
Euomphalus, 134
Euoplocephalus, 191, 192
Eupatagus, 163
Euplectella, 39
Eurypterus, 158
Eusthenopteron, 178, 199
Eutrephoceras, 142, 143
Euvola, 136, 138

F

Favosites, 117, 118
Fenestella, 123
Fissurina, 37
Flabellum, 120
Flexicalymene, 155
Fragum, 70
Furcacauda, 170, 171
Futabasaurus, 184

G

Gaojiashania, 106
Garantiana, 147, 148
Gaudryina, 37
Gemündina, 173, 174
Gerastos, 155
Gigantosaurus, 188
Ginkgoidium, 207
Ginkgoites, 207
Glenobotrydion, 102
Globotruncanella, 15
Glomospira, 58
Glossopteris, 56
Glyptocrinus, 161
Glycimeris, 136, 137
Gogia, 160
Goniatites, 144, 145
Gorgonia, 121
Grypania, 103
Gryphaea, 136, 138

Gunflintia, 98
Gyracanthus, 172, 173

H

Hadrosaurus, 190
Haigella, 84
Haikouichthys, 169, 170, 199
Halimeda, 33
Haliotis, 134
Hallopora, 122
Halysites, 117, 118
Haplophragmoides, 58
Harpoceras, 147, 148
Hastites, 150, 151
Heleosaurus, 183
Helicoplacus, 160
Heliophyllum, 66, 119
Helix, 135
Hendersonites, 72
Herrerasaurus, 186, 188
Hesperornis, 194
Hexagonaria, 119
Hexapyramis, 38
Hildoceras, 147, 148
Homoeospira, 130
Hovasaurus, 183
Huronipora, 98
Hustedia, 130
Hydnoceras 116
Hyenia, 35, 202
Hylonomus, 182
Hypagnostus, 154
Hysteroconcha, 139

I

Ichthyostega, 179
Iguanodon, 190
Inoceramus, 136, 138
Isocrania, 126
Isograptus, 48, 164, 165
Isoxys, 111

J

Jania, 33
Joannites, 145, 146
Juglans, 206
Juresania, 127

K

Kakabeckia, 98
Karpinskya, 33
Katharina, 132
Kleidionella, 122
Knorria, 202, 203
Kochiproductus, 127

L

Lacunosella, 128
Ladogoides, 128
Laevaptychus, 149
Laevidentalium, 132
Lapparia, 135
Laqueus, 41
Latimeria, 177
Lebacanthus, 175
Leonaspis, 2, 154
Leperditia, 157
Lepidocyclus, 128
Lepidodendron, 17, 202, 203
Lepidosiren, 178
Lepidostrobus, 202, 203
Leptonectes, 184
Leptotrypa, 41, 122
Leptoteuthis, 2
Levenea, 126
Limulus, 45
Linella, 98
Lingula, 126
Linotrigonia, 138, 139
Lithothamnium, 33
Lobocorallium, 119
Loboidothyris, 41
Loganella, 170, 171
Loganograptus, 164, 165
Lonchopteris, 205, 206
Lystrosaurus, 56
Lytoceras, 145, 146

M

Maclurites, 134
Macropomoides, 177
Macrostachya, 204
Manticoceras, 144, 145
Marrella, 111
Mariopteris, 205, 206
Martinia, 129
Meandrina, 41, 120
Meekopora, 122
Megateuthis, 9, 150, 151
Melanocyrillium, 103
Melonechinus, 162
Mesolimulus, 4
Mesosaurus, 56, 179, 181
Metacanthina, 155
Metalegoceras, 144, 145
Metasequoia, 17
Michelinoceras, 141
Micraster, 163
Micromelerpeton, 179, 180
Microraptor, 188, 193
Millepora, 120
Mimetaster, 23
Modiomorpha, 137, 138
Monograptus, 48, 166
Monticulipora, 122
Mucrospirifer, 129
Murex, 135
Myllokunmingia, 169
Mytilus, 136, 138

N

Nahecaris, 23
Naraoia, 111
Nautilus, 68, 142, 143
Neoceratodus, 178
Neoflabellina, 37
Neopilina, 132, 166
Neuropteris, 205, 206
Neverita, 135
Nilssonia, 207
Nipponites, 145, 147
Norella, 128
Nucula, 136, 137
Nummulites, 9

O

Ocksisporites, 36
Odontopleura, 154
Olenus, 154
Oligocarpia, 35, 205
Oncoceras, 142
Onega, 153
Orbiculoidea, 126
Orthoceras, 141
Orthostrophia, 126
Oscillatoria, 32
Osteolepis, 177, 178
Otavia, 103, 115
Ottoia, 42
Ovatoscutum, 104
Oviraptor, 189

P

Pachiteuthis, 150, 151
Pachycephalosaurus, 191
Palaeoisopus, 23
Palaeoscorpius, 23
Paleoplatoda, 104
Palmatolepis, 49
Parahesperornis, 195
Paraspirifer, 129
Parathalmanninella, 37
Paterula, 126
Pecopteris, 205
Pecten, 136
Pederpes
Peismoceras, 141
Pelagiella, 110
Peltocystis, 160
Pentacrinus, 9
Pentremites, 160, 161
Perisphinctes, 147, 148
Petalocrinus, 161
Petrolacosaurus, 182, 183
Phacops, 155
Phragmoceras, 142
Phycodes, 110
Phylloceras, 145, 146
Phyllograptus, 164, 165
Picea, 206
Pikaia, 112, 169
Pinna, 136, 138
Pisaster, 162
Pitar, 139
Placenticeras, 147, 149
Placodus, 184, 185
Plagioglypta, 132, 133
Planutenia, 110
Platyceras, 134
Plectina, 37
Plectodonta, 24
Plesiomyrmex, 18
Plesioteuthis, 150, 151
Pleurodictyum, 118
Pleuromya, 139, 140
Pleurostomella, 84

Podocarpites, 36
Pollingeria, 111
Polypora, 123
Polytilites, 156
Polyptychites, 147, 149
Posidonia, 39
Pradoia, 130
Praeplanctonia, 84
Praeradiolites, 139, 140
Primaevifilum, 96
Proclydonautilus, 142, 143
Procytheridea, 156
Proliferania, 55
Prorichthofenia, 127
Protocardia, 139
Protoceratops, 191, 193
Protoheterohelix, 84
Protopterus, 178
Protospongia, 113
Psaronius, 15
Pseudothalmanninella, 70
Pseudovoltzia, 207
Pterichthyodes, 173, 174
Pterophyllum, 19
Ptiograptus, 164
Pycnopodia, 162

R

Rabilimis, 156
Rachiosteus, 173, 174
Ramulina, 37
Raphistoma, 134
Receptaculites, 34
Remaniella, 38
Rhabdammina, 37
Rhabdinopora, 48, 164
Rhamphorhynchus, 185
Rhinodipterus, 178
Rhipidomelloides, 126
Recurvoides, 58
Repmanina, 58
Rhizammina, 58
Rhyncholites, 149
Rhynia, 202
Rusophycus, 109

S

Sacabambaspis, 170
Saichania, 191, 192
Salenia, 9
Scalarites, 145, 147
Scarburgiceras, 147, 149
Schizophoria, 126
Schloenbachia, 147, 149
Schlotheimia, 147, 148
Sclerocephalus, 179, 180
Scolithos, 3
Scutellum, 155
Sellithyris, 131
Sertella, 122
Seymouria
Sigillaria, 202, 203
Sinotubulites, 106, 117
Smilodon, 198
Sordes, 186
Spathobatis, 175
Sphagnum, 18
Spinatrypa, 130
Spirifer, 129
Spiroceras, 145, 146
Spirogyra, 32
Spirula, 150, 152
Spirulirostra, 150, 152
Spriggina, 104
Squatina, 175
Stegosaurus, 190
Stensioeina, 37
Stephanoceras, 147, 148
Stereoconus, 49
Stereosternum, 179, 181
Stethacanthus, 175
Stigmaria, 202, 203
Stringocephalus, 131
Strongylocentrous, 162
Stropheodonta, 16
Strophomena, 127
Strophonelloides, 127
Syringopora, 117, 118

T

Terebratallia, 131
Tetragramma, 9
Tetragraptus, 164, 165
Tetraspis, 154
Textilaria, 74
Thrissops, 176
Thylacosmilus, 196
Tintinopsella, 39
Tirasiana, 104
Tissotia, 147, 149
Tityosteus, 173, 199
Tonna, 135
Torridonophycus, 103
Tornoceras, 144, 145
Triceratops, 191, 193
Trichophycus, 110
Tridacna, 139
Trigonia, 139
Trigonocarpus, 206
Trionyx, 183
Triplophyllites, 119
Tropidoleptus, 126
Tubithyris, 131
Turritella, 135
Tyrannodaurus, 188

U

Uintacrinus, 161
Ullmannia, 207
Unio, 137, 138

V

Vauxia, 116
Venus, 139
Volvox, 32

W

Walchia, 207
Waldheimia, 131
Waptia, 111
Westonia, 126
Wiwaxia, 111

X

Xinjiangochara, 33

Y

Yochelcinoella, 110
Yorgia, 153
Yunnanozoon, 169

Z

Zaphrentis, 119
Zhangheotherium, 197
Zoophycos, 3

CPSIA information can be obtained
at www.ICGtesting.com
Printed in the USA
BVHW061434101122
651489BV00001B/1